Perspectives on Water Usage for Biofuels Production

Muhammad Arshad
Editor

Perspectives on Water Usage for Biofuels Production

Aquatic Contamination and Climate Change

Springer

Editor
Muhammad Arshad
University of Veterinary and Animal Sciences
Lahore
Pakistan

ISBN 978-3-319-66407-1 ISBN 978-3-319-66408-8 (eBook)
https://doi.org/10.1007/978-3-319-66408-8

Library of Congress Control Number: 2017950025

Printed on acid-free paper

This Springer imprint is published by Springer Nature
The registered company is Springer International Publishing AG
The registered company address is: Gewerbestrasse 11, 6330 Cham, Switzerland

The present book is dedicated to
MY FAMILY
A life's greatest blessing

Preface

The aim of this book is to provide texts about water-related issues arising through biofuel production, especially in response to climate changes. Major focus is on sustainable availability of clean water. It demonstrates the associations among biofuel, water and climate changes especially focusing on interdisciplinary connections that outstandingly update and enhanced understanding, deliberation and responsiveness across such disciplines.

To entice readers from different disciplines, the book offers broad perspectives on diverse aspects of climate changes impacts on water availability for present and future biofuel production. All the chapters are in-depth to highlight the relevant aspects. Major types of biofuel have been discussed to provide comprehensive compilation of relevant practical information on several aspects of biofuel, water and climate changes. The present book serves as a useful quick reference for every person involved in biofuel production or having interest in water issue and climate changes. Advanced students, researchers, instructors, decision-makers and professionals in the biofuel, water and climatic changes field will use it as a good introductory resource.

Biofuel including bioethanol, biogas and biodiesel are most promising eco-friendly substitutes to petroleum derived fuels, which are generated from renewable sources. Chapter 1 of the book illustrates biofuel, its types, applications and their feedstock resources. Further, rise of global water demand for the production of biofuel and serious outweighs greenhouse gases reduction impact of biofuels have been discussed. Chapter 2 contributes to an enhanced understanding of present climatic conditions, observed climate trends and climate vulnerability to water availability. Biofuel production processes use freshwater for different activities which becomes contaminated with organic and inorganic pollutants. Chapter 3 describes agricultural and industrial activities involving current water consumption during biofuel production. Groundwater is strategically significant due to its exceeding demand in agriculture, domestic and industrial uses. Chapter 4 keeps discussion on wastewater generation from biofuel production and how the groundwater quality is being deteriorated. In Chap 5, biofuel's effects on human health have been discussed. The water–biofuel relationship is being recognized as backbone of the factors fundamental for the future sustainable supply of biofuels. The last chapter presents prospective and future trends of the water–biofuels relationship in reference to biofuel production technologies, also the obligation of reusing wastewater and application of undiluted wastewater to grow feedstocks for biofuels to save freshwater resources.

Quick decisions must be taken now, on the use of water resources for biofuel production keeping in mind the climate change scenario to reduce the risks of future droughts and unavailability of freshwater. Less water-intensive feedstock will have to exploit if we have to avoid high-end pathways of emissions which could result in global average temperature increase. It implies that future development will increasingly need to be fuelled by less water consuming biofuel sources accompanied by much more efficient use of resources to enable development within environmental limits.

Overall, the book covers a wide range of scientific and technical aspects of water-related issues of biofuels in climate change scenario. The text is of interest to students, researchers, academicians and industrialists in the areas of water, environment, biofuel production and climate changes.

Jhang, Pakistan Muhammad Arshad, Ph.D.

Acknowledgements

I would like to send my deepest gratitude to those who have made this book possible especially the Series Editor for Springer's Water Program and all the critiques for their participation in the survey of this book and helped me to get this book written in better quality. I am also gratified to Dr. Ijaz Bano who had been a major contributing author of this book, without her willingness to leave her comfort zones and wrote about topics peripheral to her own expertise this book could not have been written. I would like to express my deep appreciation and indebtedness to Dr. Nasrullah Khan and Ms. Sidra Jamil for their support in compiling the book.

To all family, friends and others who in one way or another shared their support.

Muhammad Arshad

Contents

Contributors

Mazhar Abbas Jhang-Campus, University of Veterinary and Animal Sciences, Lahore, Pakistan

Mushtaq Ahmad Department of Plant Sciences, Quaid-I-Azam University, Islamabad, Pakistan

Muhammad Arshad Jhang-Campus, University of Veterinary and Animal Sciences, Lahore, Pakistan

Ijaz Bano Jhang-Campus, University of Veterinary and Animal Sciences, Lahore, Pakistan

Ammanullah Khan Jhang-Campus, University of Veterinary and Animal Sciences, Lahore, Pakistan

Abdur Rahman Jhang-Campus, University of Veterinary and Animal Sciences, Lahore, Pakistan

Farman Ali Shah Department of Chemical Engineering, Mehran University of Engineering and Technology, Jamshoro, Pakistan

Muhammad Younus Jhang-Campus, University of Veterinary and Animal Sciences, Lahore, Pakistan

Muhammad Anjum Zia Department of Biochemistry, University of Agriculture, Faisalabad, Pakistan

Chapter 1
An Overview of Biofuel

Muhammad Arshad, Muhammad Anjum Zia, Farman Ali Shah and Mushtaq Ahmad

Abstract Fossil fuels applications are linked with current widely held environmental issues. The decline of these fuels resources with environmental penalties has compelled for substitutes and usage of renewable biofuel as energy sources; has gained a significant importance in last two decades. Production of biodiesel, biogas and bioethanol from various feedstock, several kinds of wastes, many types of biomass and agricultural residues, is ecological viable and sustainable option. The involvement of biofuel in worldwide transportation fuels seems to be revolving about 5% over the next decade. But, many studies put forward that biofuel may share up to a one fourth of transport fuel supplies by 2050. In the first part of the chapter, advantages and applications of mostly used biofuel is presented. The second part of the chapter keeps concepts about biodiesel. Biogas production and composition has been addressed in third portion. Finally, the production of bioethanol from different feedstock has been discussed. Instability of fossil fuels prices in last decade and environment concerns has increased biofuel production many folds. Such a fast growth has been resulted controversial and raised some concerns over potential water use in production of biofuel.

Keywords Global warming · Biofuel · Bioethanol · Biodiesel · Biogas

M. Arshad (✉)
Jhang-Campus, University of Veterinary and Animal Sciences, Lahore, Pakistan
e-mail: Muhammad.arshad@uvas.edu.pk

M.A. Zia
Department of Biochemistry, University of Agriculture, Faisalabad, Pakistan

F.A. Shah
Department of Chemical Engineering, Mehran University of Engineering and Technology, Jamshoro, Pakistan

M. Ahmad
Department of Plant Sciences, Quaid-I-Azam University, Islamabad 45320, Pakistan

M. Arshad (ed.), *Perspectives on Water Usage for Biofuels Production*,
https://doi.org/10.1007/978-3-319-66408-8_1

1.1 Introduction

Presently, three major issues are in front of human beings: hunger, the lack of energy and the deterioration of the environment (Popp et al. 2014). It is obligatory to fight with all the three vehemence simultaneously, because any one of these is capable to extinct out our civilization (Escobar et al. 2009). The ease of access to energy is the basic driving force behind the socio-economic progress and the vital element to sustain human's current elevated standard of living (Walker et al. 2016; Arshad and Ahmed 2016). Globally consumption of energy has been almost doubled up in recent times (Bentley 2016) and fossil fuels share more than 80% (Pfenninger and Keirstead 2015). When we talk about energy, it is clear to each and every person that its saving is the best attitude to be privileged by minimizing irrational use and enhancing the utilization efficiency (Abdmouleh et al. 2015) as fossil fuel reservoirs are depleting fast (Hook et al. 2014). Up till now human's energy requirements has been met by the fossil fuels (coal; oil; gas) since many decades. Alternative cheap and environment friendly energy is the hot issue in today's world. Fossil fuels account for over 80.3% of the primary energy consumed in the world, and 57.7% of that amount is used in the transport sector (Escobar et al. 2009). Burning of conventional fuels results in the harmful emissions of greenhouse gases such as carbon dioxide (CO_2), nitrogen oxide (NO_x), volatile organic compounds (VOC) and hydrocarbons (HC); incremental for the climate changes (Chavez-Baeza and Sheinbaum-Pardo 2014; Friedlingstein et al. 2014; Reuter et al. 2014). Although such fuels give the best cost/benefit ratio; but at the same damage the environment. Fossil based diesel is an essential fuel for running vehicles, power plants and motor engines in the transportation, agricultural and industrial sectors (Emanuel and Gomes 2014; Orsi et al. 2016) and remained the most merchandising commodity among primary products trade in 2010 (Janaun and Ellis 2010). Transportation sector spent more than 30% of the energy supply globally, in which above 80% is by the road transport (Holmberg and Erdemir 2015). Worldwide almost 60% oil supply is consumed by this sector (Bilgen 2014), practically operating on gasoline, diesel oil almost 97.6%, with a small amount from liquid natural gas (Ramadhas et al. 2004; Murphy et al. 2013).

Now the world has been challenged by global warming problem (IPCC 2014). The release of Carbon dioxide from the combustion of fossil fuels, the key contributor to the process has generated the interest in promoting biofuel as one of the leading renewable energy sources (Kumar et al. 2013). Table 1.1 shows the major benefits of the biofuel. The sustainable production of biofuel is a valuable tool in stemming climate change (Creutzig et al. 2015), boosting local economies, particularly in lesser-developed parts of the world (van Eijck et al. 2014a; van Eijck et al. 2014b), and enhancing energy security for all (Jatrofuels 2012; Hughes et al. 2014). Advancement in renewable biofuel sources; cling to solution key of the dual difficulties, running down the fossil fuel reservoirs and environmental pollution (Smith 2013). Therefore, exploration of novel, renewable, environment friendly, clean, reliable and economically feasible energy resources is serious requisite of the

Table 1.1 Major benefits of biofuels (Balat 2011a, b)

Commercial value	Variety in fuel mix More sustainable Ability to create many rural jobs Can increased the government revenue through taxes Industrial investments (plant and equipment) will be increased Farming/agricultural sector can be developed Less international competition Independence from imported petroleum
Climate change effects	Reduction in release of greenhouse gases Air pollution can be minimized Easy for biodegradation Better combustion efficiency Better carbon sequestration
Indigenous impacts	To achieve the domestic targets More reliability in supply Reduced utilization of fossil oils Ready availability Indigenous distribution

day (Dale et al. 2014). The discharge of greenhouse gases through the burning of fossil fuels in transport sector alters the natural equilibrium of environment. The world has now started to realize the problem and syndromes created by conventional fuels (Karwat et al. 2014). To minimize the fossil fuels role, the exploration of renewable substitutes, the biofuel like bioethanol, biogas and biodiesel are on rise (Ho et al. 2014). Biofuel, biodiesel, biogas and bioethanol are currently available in the market, already being used for various types of engines (Prasad et al. 2007; Demirbas 2008; Janaun and Ellis 2010; Shahid and Jamal 2011; Geczi et al. 2015; Malakhova et al. 2015; Choudhary et al. 2017).

The "bio" in biofuel refers to crop and wood-based raw materials such as molasses, rice husks, corn and wood waste, which are processed into fuel. For developed countries, biofuel offer prospects for meeting their emission reduction commitments under the Kyoto Protocol (de Alegría et al. 2016). For developing countries, biofuel present a means to both reduce energy import bills as well as earn precious foreign exchange (Khan 2007). Biofuel are produced from bio-origin resources by thermochemical processes (Liu et al. 2008; Balat et al. 2009; De Kam et al. 2009; Alonso et al. 2010; Sims et al. 2010; Ertas and Alma 2011) and biochemical process (Amin 2009; Uddin et al. 2016; Shukla et al. 2017). Biomass is reformed in thermo-chemical catalytic and non-catalytic processes such as pyrolysis, gasification, liquefaction, supercritical fluid extraction, supercritical water liquefaction to produce maximum energetic exploitable liquid and gaseous products. Biofuel include bioethanol, biodiesel and biogas produced through biochemical path ways such as, alcoholic fermentation, anaerobic fermentation and trans-esterification (Balat 2011a). Trans-esterification of vegetable oils, animal fats, waste oils/fats, used oils/fats and microbial oils with methanol and to some extent with ethanol/butanol results in biodiesel production (Hoekman et al. 2012; Arshad et al. 2014a). In spite of the constant requirement of biodiesel production, the lack of oil's feedstock have become problematic and decrease the production of first generation biodiesel as there was an almost two-fold increase in the price of conventional plant oils (Choi et al. 2010). The reply was production of biodiesel from "non-conventional oils" and with the aid of microorganisms capable of producing intracellular lipids (Koberg et al. 2011) called as second generation biodiesel. The third generation biodiesel is derived through atmospheric CO_2 sequestration (Thiyagarajan et al. 2017; Bhola et al. 2014).

Worldwide ethanol production is based upon petrochemical and biochemical methods (Van Uytvanck et al. 2014). In the petrochemical method ethylene is hydrated in the presence of mineral acids (Ren et al. 2015). The process is much attractive, if the price of raw material remains low. But price comparison between ethanol and ethylene, has displaced the method almost completely by the processes depending on the treatment of biomass (Haro et al. 2013). Ethanol fermentation of the glucose is the oldest technique and also used to produce alcoholic beverages (Majchrowicz 2013). Agricultural based stuffs containing sugar, starch, and cellulose are employed as raw material (Rothman et al. 2015). Normally the fermentation process by the yeasts occurs at room temperature, anaerobically (Mielenz 2014). The general equation of the process is represented as,

$$C_6H_{12}O_6 \rightarrow 2C_2H_5OH + 2CO_2$$

There arise two molecules of CO_2 and ethanol for each molecule of the glucose fermented (Sarris and Papanikolaou 2016). Industrial alcoholic fermentation process normally halts as ethanol concentration approaches at 9–10% (Dai et al. 2014) participating yeasts can be used in subsequent cycles of fermentation (Stanbury et al. 2013). The ethanol yield from glucose is 88–95% (Luterbacher et al. 2014) with some byproducts such as glycerin (3–5%) and acetic acid etc. (Onuki et al. 2016). Fermented mash is distilled (Borse and Sheth 2017) to increase the ethanol concentration up to 94% (Mayer et al. 2015), bring it to the required marketability (Duffield et al. 2015). Resultant bioethanol can be further purified through dehydration process (Vázquez-Ojeda et al. 2013). Bioethanol in both forms hydrated and dehydrated (also called absolute alcohol) is used as fuel in pure or blended with gasoline (Foong et al. 2014). The production of bioethanol is accompanied by serious economic and environmental benefits (Maroun and La Rovere 2014), since ethanol as a fuel presents a high octane number (Leone et al. 2014), while even small amounts of ethanol added into the gasoline can significantly increase the octane number of the blend (Foong et al. 2014). Moreover, the higher oxygen content improves the efficiency of the combustion (Rakopoulos et al. 2014). Also green house gas emissions are generally considered to be reduced as ethanol burns results in lower emission of carbon monoxide (CO), volatile organic compounds, sulfur oxides, etc. in comparison with the burn of the typical fossil fuels (Dwivedi et al. 2015).

Biogas is another mostly used fuel and its significance as renewal biofuel is well recognized (Lee et al. 2014). It is produced during anaerobic digestion of biodegradable organic materials (Ariunbaatar et al. 2014) and typically keeps approximately 60–65% methane (Divya et al. 2015). It can be used to offset part of the energy requirement (Rahman et al. 2014). Biogas production, collection and utilization methods have been gradually competent to improve the quality (Havukainen et al. 2014). Energy recovery from biogas is developing into a successful "waste/residues to bioenergy" technology (Gonzalez-Salazar et al. 2016). Biogas is commonly used as fuel in the boilers, employed in combined heat and power applications to electricity generation and to make steam (Wellinger et al. 2013; Poschl et al. 2010). Overall efficiency of biogas use can approach 80% if all the recovered heat is used. The quantity and quality of gas produced during anaerobic digestion depends on the feed characteristics. Several methods are available to estimate methane generation from a waste stream during anaerobic digestion. Knowing the chemical composition of the waste stream, the methane production can be estimated (El-Mashad and Zhang 2010).

1.2 Global Interest in Biofuel

Biofuel have been well thought-out as supplement to fossil fuels for transportation since the oil crises of 1973 and 1979 (Kalam and Masjuki 2002). Attention toward biofuel has been resurged in the early 2000s due to heavy distresses about climate change, depleting fossil oil reserves with fluctuations in price (Fogel 2007). More than 40 countries have formulated their national policies to sustenance the biofuel (Timilsina 2014). Table 1.2 shows the list of the countries that have set their targets for the biofuel. Current fascinating biofuel production has burst an aggressive talk, whether to support the policies and programs about biofuel production or not. Because at one end the biofuel are promoted as a solution to climate change issues and setting up better energy supply globally (Headey and Fan 2008; Tilman et al. 2009; Lynd and Woods 2011) while, at the other end, it is indicated that these are a risk to food supply with stress on water supply on earth (Diouf 2007; Pimentel et al. 2009; Borras et al. 2010). Claims for reduction in green house gases discharge, (the basic argue to support biofuel) have also been challenged (Searchinger et al. 2008; Danielsen et al. 2009). Such discussions have lessened the earlier enthusiastic support to biofuel.

1.2.1 Different Types of Biofuel Today

From a long list of biofuel only biodiesel, bioethanol and biogas are presently produced as a fuel on an industrial scale (Antoni et al. 2007). These fuels make up to more than 90% of the biofuel market (Demirbas 2009c, d). All biofuel have to exhibit defined chemical and physical properties, meeting the demands of engine application such as stability and predictable combustion at high pressures as well as the demands of transportation such as safety and energy density. Table 1.3 shows

Table 1.2 Liquid biofuels mandates and targets of selected countries globally (Arshad 2010)

Country	Ethanol	Biodiesel
Australia	E6	B2
Argentina	E5	B7
Belgium	E4	B4
Bolivia	E10	B20 by 2015
Brazil	E18-E25	B5
Canada	E5	B2
China	E10	
Colombia	E10	B20
Costa Rica	E7	B20
Dominican Republic	E15	B2
EU	10% renewable in transport (2020)	
Ethiopia	E10	
Finland	E5.75	B5.75
Germany	E10	
India	E5; E20 (2017)	B20 (2017)
Indonesia	E15	B20
Italy	E5	B5
Jamaica	E10	B5
Malaysia		B5
Malawi	E20	
Mozambique	E10	B5
Te Netherland	E4	B4
Norway		B3.5
Pakistan		B5
Panama	E10	
Paraguay	E18-E24	B5
Peru	E7.8	B5
Philippines	E10	B2
South Africa	2% of transport energy	
South Korea		B2.5
Thailand	E10	B3

various gaseous and liquid biofuel. Liquid biofuel can be stored, distributed, carried and used as an energy source in cars, trucks, trains and planes without any difficulty (Nigam and Singh 2011). Biogas is gaseous in nature and somewhat difficult to transport. It requires a separate distribution infrastructure to be developed.

Liquid biofuel have to remain in a liquid state and pumpable at all temperatures encountered. Further requirements on liquid biofuel are a high heat of combustion value to reduce energy losses and costs during transportation and stability during storage. Some longer chain alcohols like butanol have a heat of combustion sufficiently high to allow for their use in high thrust-to-weight applications such as

Table 1.3 The biofuels with possible production route, their use and the applications to engines (Antoni et al. 2007)

Biofuel	Process	Status	Engine application
Biomethanol	Thermochemical/microbial	Pilot plant	[pure/blend] MTBE/biodiesel
Bioethanol	Microbial	Industrial	Pure/blend
Biobutanol	Microbial	Pilot plant/Industrial	Pure/blend
ETBE	Chemical/Microbial	Industrial	blend
Biomethane	Microbial	Industrial	Pure/blend
Biohydrogen	Microbial	Laboratory	Pure/blend
Pure biodiesel	Physical/chemical (enzymatic)	Industrial (laboratory)	Pure/blend

airplanes (Yanai et al. 2015). Safe and environment friendly storage, vapour pressure and ignition temperature are important factors.

1.2.2 Economics of Biofuel

The economics of biofuel is majorly determined by the value of the feedstock used for their production (Elbehri et al. 2013). For first-generation biofuel raw material price accounts approximately between 60 and 90% of the total production (Ho et al. 2014; Tan et al. 2013). The price competitiveness of biofuel to petroleum counterparts varies between countries and with the feedstock used (Wesseler and Drabik 2016). The "factory gate" price of Brazilian ethanol remained lesser than the "refinery gate" price of gasoline in last decade (Onal and Nunez 2014). Both Brazilian and US ethanol remains expensive than gasoline on an energy equivalent basis. Sugarcane derived Brazilian ethanol is better competitive than US ethanol, but is still usually more expensive than gasoline. In case of biodiesel, it is more expensive than diesel, even though a liter of biodiesel provides around 14% less mileage than diesel. While the biogas is much completive due to unavailability of natural gas everywhere (Alam and Hasan 2017).

Brazilian ethanol production cost remains lower than for US corn or European wheat ethanol, due to use of sugarcane bagasse in boilers; to come across on site steam and power demand. Moreover the production of biogas and its utilization in electricity can further lower the production cost. Ethanol production cost from wheat grains can be lowered if the impact value of by-products is considered. Likewise, biodiesel production cost will be fall, if main byproduct, glycerin can fetch a market value, which is utilized in the beverages, food and pharmaceuticals industries (Jonker et al. 2015; Losordo et al. 2016). To produce ethanol from sugarcane molasses is cheaper than from sugarcane itself (Castañeda-Ayarza and

Cortez 2016). Biodiesel from non-food seeds like jatropha largess an interesting, alternative if yields can be improved to commercial level sand if sufficient low-cost labour can be assembled for the highly labour intensive seed collection process (Carriquiry et al. 2011). The capital costs of second-generation biofuel account for a higher portion, while the feedstock costs are significantly lower as compared to first-generation biofuel (van Eijck et al. 2014). Overall production costs from micro-algae appear to be higher presently, but could fall in the future as well as technology improves and production expands (Kern et al. 2017).

Biofuel, like fossil fuels, come in a number of forms and meet a number of different energy needs. The present book chapter is an introduction of the major globally used biofuel, biodiesel, bioethanol and biogas. Each type of the biofuel has been explained well.

1.3 Biodiesel

Biodiesel is a mono-alkyl ester of fatty acids from vegetable oil and is presently produced by catalytically trans-esterification process with petro-chemically derived methanol (Ma and Hanna 1999). The glycerol produced during trans-esterification creates a deposit problem in some areas. It could be fermented, e.g. to 1,3-propanediol and possibly to other products by metabolically engineered bacteria (Calero et al. 2015) or to methane in biogas plants where it can be added in low concentrations as co-substrate. Instead of using vegetable oil, microalgae could be grown in photo-bioreactors for the production of suitable oil. Because of their high oil productivity, the specific demand of land area needed is strongly reduced by this concept in contrast to oil from plants (Chisti 2007). Mono-alkyl esters of long chain fatty acids originated from renewable lipid sources such as plant oils, animal fats or algal sources for use in compression ignition (diesel) engines" are called as biodiesel (Kafuku and Mbarawa 2010; Satyanarayana and Muraleedharan 2011; Shahid and Jamal 2011; Ghazali et al. 2015). Biodiesel can be superior replacer of conventional diesel as it based on oxygenated esters of long chain fatty (Ong et al. 2011; Hoekman et al. 2012).

Global warming issue can be well managed with the use of biodiesel in transportation sector. Biodiesel is highly biodegradable (Hossain and AlEissa 2016) and has minimal toxicity and its ignition in diesel engines can withdraws the total unburned hydrocarbons (HC) and polycyclic aromatic hydrocarbons (PAHs) (Hoekman and Robbins 2012). İt can replace diesel fuel use in boilers and internal combustion engines without major modifications (Bergthorson and Thomson 2015) and can significantly reduce the particulate matter and carbon monoxide emission (Basha et al. 2009). Emissions of sulphates, aromatic compounds and other chemical substances, that are destructive to the environment are almost zero (Popovicheva et al. 2015). High flash point, better lubrication, and high Ocetane number with very close physical and chemical characteristics to those of fossil diesel (Rashedul et al. 2015) allow its application as pure biodiesel, B100 or may be blended with fossil diesel fuel with minute technical adjustments (Basumatary 2015). Wolrdwide many countries such as Malaysia, United States of America, Brazil, Germany and many other European states are using it due to its potential for better safeguard the environment from hazards emissions and protect the human health from potential or probable threats (Canakci et al. 2009; Cremonez et al. 2015; Johari et al. 2015; Knothe et al. 2015; Eryilmaz et al. 2016; Tsoutsos et al. 2016).

1.3.1 Biodiesel Feedstock

The wide range of feedstock for biodiesel production is available to sustain newly emerging biodiesel industry. The availability of feedstock depends upon some factors such as climate of the region, geographical locations, soil conditions and agricultural practices of a country. Worldwide, more than 350 oil crops have been identified as potential sources of biodiesel production (Bart et al. 2010). Presence of oil percent and the yield per hectare are important parameters in feestock selection. (Atabani et al. 2012; Tabatabaei et al. 2015) have reported estimated oil content and yields of many biodiesel feedstock. Oil composition, type and ratio of fatty acids present impact the fitness of oil as a raw material for biodiesel production (Basumatary 2015). Alone feedstock denote 75% of the overall biodiesel production cost (Ahmad et al. 2011; Atabaniet al. 2012). So, selection of the cheapest feedstock is vital to reduce cost of biodiesel production cost. Generally, feedstock of biodiesel production are classified into four major categories (Satyanarayana and Muraleedharan 2011):

A. Edible plant oils
B. Non-edible plant oils
C. Waste or recycled oils
D. Animal and poultry fats

Some non-edible and edible oil sources used for biodiesel production have been shown in Table 1.4. Applications of edible oils have generated lots of concerns such as food versus fuel debate, creation of serious environmental problems such as grave destruction to soils, deforestation and consumption of arable land/water (Balat and Balat 2010; Balat 2011b; Deng et al. 2011). Edible oils become un-feasible in the long term because of the expected growing gap between supply and demand of such oils (Chapagain et al. 2009).

Non-edible oils can reduce the utilization of the edible oil for biodiesel production. Non-edible oil resources are easily available in many parts of the world especially wastelands that are not suitable for food crops, eliminate competition for food, reduce deforestation rate, more environmentally friend, produce useful by-products and they are very economical as compared to edible oils (Sarma et al. 2005; Chhetri et al. 2008; Gui et al. 2008; Murugesan et al. 2009; Sarin et al. 2009;

Table 1.4 Some non-edible and edible oil sources used for biodiesel production

Source	Oil yield (kg oil/ha)	Oil yield (wt%)	References
Non-edible oil			
Jatropha	1590	Seed: 35–40 Kernel: 50–60	Gui et al. (2008)
Rubber seed	80–120	40–50	Ramadhas et al. (2005)
Castor	1188	53	Saka (2005)
Pongamiapinnata	225–2250	30–40	Karmee and Chadha (2005)
Edible oil			
Soybean	375	20	Gui et al. (2008)
Palm	5000	20	Berchmans and Hirata (2008)
Rapeseed	1000	37–50	Westbrook et al. (2011)

Table 1.5 Some animal sources of fat and their fatty acid compositions

% (By weight)	Beef tallow[a]	Chicken fat[b]	Pork lard[c]	Mutton fat[d]
Lauric acid (C12:0)	–	–	–	0.2
Myristic acid (C14:0)	2.72	0.5	1.7	3
Palmitic acid (C16:0)	25.3	24	23.2	27
Palmitoleic acid (C16:1)	2.02	5.8	2.7	2
Stearic acid (C18:0)	34.7	5.8	10.4	24.1
Oleic acid (C18:1)	29.87	38.2	42.8	40.7
Linoleic acid (C18:2)	0.75	23.8	19.1	2
Linolenic acid (C18:3)	–	1.9	64.7	–

[a]Ma and Hanna (1999), [b]Wyatt et al. (2005), [c]Dias et al. (2008), [d]Mutreja et al. (2011)

Saravanan et al. 2010; Falasca et al. 2010; Kumar and Sharma 2011; Atabani et al. 2012; Banković-Ilić et al. 2012; Mofijur et al. 2013; Shirazi, et al. 2014; Haile 2014; Zhang et al. 2015). Table 1.5 shows some fat sources of animal origion with relevant fatty acid composition.

Microalgae have emerged as third generation biodiesel feedstock. These are photosynthetic microbes capable to convert sunlight, water and CO_2 to algal biomass, more efficiently as compared to conventional crops. High oil content, better growth rates and productivity as compared to edible and non-edible feedstock make microalgae as promising feedstock. Table 1.6 shows micro algal strains keeping various quantities of oils that can be further processed to produce biodiesel. Up to 25 times higher yields than oil palm and 250 times than soybeans is achieved through the algal cultivations (Sharma and Singh 2009; Singh and Singh 2010; Ahmad et al. 2011).

Table 1.6 Some microalgal strains capable for biodiesel production keeping oil contents (% dry wt)

Microalga	Oil content	References
Botryococcus braunii	25–75	Metzger and Largeau (2005)
Chlorella sp.	28–32	Li et al. (2015)
Crypthecodiniumcohnii	20	Brennan and Owende (2010)
Cylindrotheca sp.	16–37	Meng et al. (2009)
Isochrysis sp.	25–33	Renaud et al. (1991)
Monallanthussalina N	20	Mata et al. (2010)
Nannochloris sp.	25–35	Gouveia and Oliveira (2009)
Nannochloropsis sp.	31–68	Brown et al. (2010)
Neochlorisoleoabundans	35–54	Popovich et al. (2012)
Nitzschia sp.	45–47	Demirbas and Demirbas (2011)
Schizochytrium sp.	50–77	Ratledge (2004)

1.3.2 Biodiesel Production Technologies

In last decade, biodiesel production is gone through fast technological improvements in industries and academia. Higher production cost is the major drawback in its commercialization. Various studies on the economic improvement of technologies and methods have been conducted in search of optimal conditions of biodiesel production. The primary methods to make biodiesel are direct use and blending of vegetable oils, micro-emulsions, thermal cracking (pyrolysis) and trans-esterification (Ma and Hanna 1999). Trans-esterification process is the most common method used in the biodiesel industry, in which vegetable oil or animal fat and an alcohol (methanol, ethanol) react in the presence of a catalyst or without the use of catalysts (Demirbas 2009a).

1.3.2.1 Direct Use and Blending of Oils

Application of vegetable oils as fuels stems around since 1900 when Dr. Rudolph Diesel, firstly experienced Peanuts oil in his newly invented compression engine. The direct applications of vegetable oils as fuel are problematic and have many weak spots in diesel engines. Although it's being researched comprehensively for the previous few decades, but experimentation started for about hundred years. Vegetable oils may be blended with diesel fuels to better the viscosity so as to make solution of the problems linked with the use of pure vegetable oils (Koh and Ghazi 2011). Blending ratios of 1:10 to 2:10 vegetable oil to diesel fuel have been found to be better rather than direct use of vegetable oils. Increased thickness due to high viscosity, presence of acid components, higher free fatty acids ratio, with the gum formation are some apparent teething troubles (Ma and Hanna 1999).

1.3.2.2 Micro-Emulsion of Oils

The formation of microemulsions is a potential solution for resolving the problematic high vegetable oil viscosity issue. A colloidal equilibrium dispersion that is clear, stable with three components: an oil phase, an aqueous phase and a surfactant, of optically isotropic fluid microstructures with dimensions generally in the 1–150 nm range formed spontaneously from two normally immiscible liquids and one or more ionic or non-ionic amphiphiles (Fernandes et al. 2012) is called micro-emulsion. Such fuels are also called "hybrid fuels" (Satyanarayana and Muraleedharan 2009).

The solvents like methanol, ethanol and 1-butanol have been studied and micro-emulsions with butanol, hexanol and octanol can meet the maximum viscosity limitation for diesel engines (Oner and Altun 2009).

1.3.2.3 Pyrolysis of Oils

Conversion of one organic compound into some other substance using heat with or without presence of a catalyst is called pyrolysis. Vegetable oil, animal fats, natural fatty acids or methyl esters of fatty acids can be subjected to pyrolysis (Yusuf et al. 2011). Thermal cracking of triacylglycerol's is much promising method for biodiesel production as it is very similar to petroleum refining (Maher and Bressler 2007). Liquid product fractions resulted through thermal decomposition of vegetable oils are closely approaching to characteristics of fossil diesel oil and reported as suitable for diesel engines. Pyrolysis process can further divided into catalytic and non-catalytic processes (Leung et al. 2010). Equipment/machinery used for pyrolysis and thermal cracking is much expensive (Ma and Hanna 1999).

1.3.2.4 Trans-Esterification of Oils

The most used method for biodiesel production is trans-esterification of oils with alcohol (methanol or butanol). Glycerin is major byproduct of this reaction. In the first step, the triglycerides are changed into diglycerides, and diglycerides are converted to monoglycerides and glycerol, yielding one methyl ester molecule from each glyceride at each step (Ma and Hanna 1999). Most important process variables are temperature, time, proportion of alcohol to oil, catalyst concentration, mixing force (RPM) and type of feedstock used (Marchetti et al. 2007). As alcohols and triglycerides are immiscible to generate a mixture of single phase, therefore surface contact between these two reactants remains very low and causes the trans-esterification reaction to proceed relatively slow. Presence of a catalysts makes the surface contact better among the reactants; thus speed-up the reaction. Henceforth, the researchers have been exploring alternatives that can solve the problems (Demirbas 2009b).

Catalytic biodiesel production The oils are transesterified by warming them with an alcohol and a catalyst. If the catalyst remains in the same phase in which reactants (liquid phase) throughout trans-esterification process, it is called as homogeneous catalyst. When the catalyst remains in different phase to that of the reactants, then it is called as heterogeneous catalyst (Zabeti et al. 2009). Application of the appropriate catalyst is the vital to lower the biodiesel production cost. Presently, commercial biodiesel is prepared by using homogenous catalyst. (Ragit et al. 2011).

Homogeneous catalytic transestrification The homogenous catalysts used in transesterification reactions are classified into basic and acidic catalysts. Basic type catalysts used transestrification processes needs a very high purity of raw materials with post reaction separation of catalyst, byproduct, and product. These conditions increase the cost of biodiesel.

Heterogeneous catalytic trans-esterification Heterogeneous catalysts act in a different phase from the reaction mixture in such type of trans-esterification. The catalysts can be easily separate and reuse. Moreover use of heterogeneous catalyst does not yield soap (Leu 2013). The heterogeneous catalytic systems of trans-esterification put forward the exclusion of different steps like washing, separation of biodiesel and catalyst. Higher efficiency with better profitability is the key features of the process.

Non-catalytic biodiesel production There are only two trans-esterification processes in which no catalyst is employed. These are supercritical alcohol process and BIOX process.

Supercritical alcohol trans-esterification In supercritical alcohol method, instead of using catalysts, high pressure and temperature are applied to do the trans-esterification reaction. Reaction becomes faster and conversion just occurs in (50–95%) the first 10 min. The required temperature ranges from 250 to 400 °C (Meher et al. 2006; Teo et al. 2014).

BIOX co-solvent trans-esterification As the oils are not well soluble in alcohols, so the rate of trans-esterification remains very slow. To solve the issue, another tactic is being used in form of co-solvent which can solve both. Tetrahydrofuran, has a boiling point much closer to methanol can solve the issue (Kusdiana and Saka 2004).

1.3.3 Microdiesel

E. coli cells were metabolically engineered by introducing the pyruvate decarboxylase and alcohol dehydrogenase genes pdc and adhB, respectively, from *Zymomonas mobilis* for abundant ethanol production. The gene atfA for an unspecific acyltransferase from *Acinetobacter baylyi* was introduced to esterify ethanol with the acyl moieties of CoA thioesters of fatty acids. If the cells are grown aerobically in the presence of glucose as an energy and carbon source and of oleic acid, ethyl oleate was the major product. However, de novo synthesized fatty acids were not used by the acyl transferase, which made the external addition of fatty acids necessary. This indicates that considerable further development is needed. However, a new concept of the microbiological production of biodiesel has been shown with these experiments (Barney 2014).

Conversion of plant oil to biodiesel is a mature technology. However, microbial contribution to the production process is close to zero at present. Inclusion of biologically fermented ethanol and butanol will not pose technical problems. The use of enzymes or biological systems in trans-esterification is to be developed. Most diesel cars are now licensed to use a biodiesel diesel blend of up to 5% (v/v). The conversion of a conventional diesel engine for pure biodiesel use is offered by many companies and costs in Germany up to 1,500€ per car. The modified engine, however, requires more frequent engine oil changes.

1.4 Biogas

The world is progressively more looking for the imperative nature of sustainable development due to environmental concerns caused by the burning of fossil fuels. Therefore joint research on energy and environment is growing day by day, in both R&D and technology implementation level. Microbes convert organic matter into biogas through a natural process called anaerobic digestion. The process naturally occurs in marshes, landfills, wetlands, and also in the digestive tract of ruminants. It is quite possible to collect biogas and can easily be utilized as an energy resource. It can also yield valuable industrial products or byproducts. The value of biogas has been risen up due to two causes: (i) for the reason that its liberation into the atmosphere contributes principally to increase greenhouse gas volume (ii) its energetic contents are high, so those make it valuable.

Biogas plants produce methane gas sustainably along with carbon dioxide from plant biomass, which may come from organic household or industrial waste or from specially grown energy plants (Divya et al. 2015; Mao et al. 2015). The general composition of biogas and value of its components has been shown in Table 1.7. The advantage of the biogas process is the option to use the polysaccharide constituents of plant material to produce energy, such as electrical power and heat, in relatively easy-to-manage and small industrial units. Alternatively, the gas can be compressed after purification and enrichment and then fed to the gas grid or used as a fuel in combustion engines or cars.

Its greatest advantage is the environmentally friend aspect of the technology, which includes the potential for complete recycling of minerals, nutrients (phosphate etc.) and fiber material (for humification), which come from the fields and return to the soil, playing a functional role by sustaining the soil's vitality for future plantation. The technology is currently mature, but there is plenty of room for optimization, which will result in large high-tech production plants with integrated utilization of by-products.

Table 1.7 General characteristics of the biogas produced through anaerobic digestion

Characteristics	Corresponding values	References
Composition	55–70% methane (CH_4) 30–45% carbon dioxide (CO_2) Traces of other gases	Rasi et al. (2007)
Energy content	6.0–6.5 kWh m^{-3}	Rao et al. (2010)
Explosion limits	6–12% biogas in air	Kapdi et al. (2005)
Ignition temperature	650–750 °C	Kolbitsch et al. (2008)
Critical pressure	75–89 bar	Kapdi et al. (2005)
Critical temperature	−82.5 °C	Kapdi et al. (2005)
Normal density	1.2 kg m^{-3}	Esteves et al. (2008)
Smell	Rotten eggs smell	Rasi et al. (2007)
Molar mass	16.043 kg $kmol^{-1}$	Esteves et al. (2008)

1.4.1 Substrates for Biogas Production

Generally, biomass of any kind containing carbohydrates (Starch, cellulose and hemicellulose), proteins and fats as core components can be employed for biogas production. Following points are important for the selection of biomass for biogas production (De Francisci et al. 2014).

- The composition of organic matter has to carefully chosen for fermentation process.
- The potential of the organic matter for biogas formation should be as high as possible.
- Selected substrate must be free from pathogens and other microbes which can harm the fermentation process.
- Biogas composition has to be examined proper for further use.
- The fermentation residues may keep the suitable content to be applied it as fertilizer.

A lot of substrates have been used for biogas production and reported in the literature. In Table 1.8 a comprehensive list of different substrates utilized for biogas production has been provided.

Cow manure is better substrate and is also useful for inoculation, manure from other farm animals such as pigs, chickens and horses, fat from slaughter waste or frying oil, organic household or garden waste, municipal solid waste and rotten foodstuff is equally applicable for anaerobic digestion. Even organic waste from hospitals containing paper and cotton, municipal sewage sludge, waste from agriculture or food production, organic-rich industrial waste water etc. can be used as consumable substrate. Often, energy crops such as maize (whole plant including the corn), clover, grass, young poplar and willow are especially grown for biogas production and added purely or in mixture. To ensure a homogeneous substrate quality throughout the year, the green plant material is usually stored as silage,

Table 1.8 Different feed stocks used for biogas production reported in literature

Substrate for biogas production	References
Residuals from beverage production	
Spent grain, fresh or ensilaged	Malakhova et al. (2015), Wolters et al. (2016)
Spent grain, dry	
Apples pulp	Géczi et al. (2015)
Apple mash	Kafle and Kim (2013)
From fruits and vegetable waste	Sagagi et al. (2009)
Animal waste	
Slaughterhouse waste	Ware and Power (2016), Fathya et al. (2014)
Meat and bone meal	Zarkadas et al. (2015)
Fat from the separator used ingelatine production	Moeller and Görsch (2015)
Animal fat	Martínez et al. (2016)
Blood	Abdeshahian et al.(2016)
Greens, grass, cereals, vegetable wastes	
Vegetable wastes	Scano et al. (2014), Janczak et al. (2016)
Grass	Rodriguez et al. (2017)
Hay	Zhu et al. (2014)
Meadow grass, clover	Kristensen et al. (2016)
Market wastes	Sridevi et al. (2015)
Leaves of sugar beet	Ohuchi et al. (2015)
Wheat bran	Wolters et al. (2016)
Soybean	Zhu et al. (2014)
Giant cane, cornsilages and pig slurry	Luca et al. (2015)
Sugar beet cossettes and pig manure	Aboudi et al. (2015)
Olivepomace and milk whey	Battista et al. (2015)
Food waste and rice husk	Haider et al. (2015)
Many flower, silvergrass and microalgae	Li et al. (2015)
Sugar beet pulp silage and vinasse	Zieminski and Kowalska-Wentel (2015)
Cow slurry, apple pulp and olive pomace	Riggio et al. (2015)
Food waste and cattle manure	Zarkadas et al. (2015)
Rice straw and cow manure	Li et al. (2015a)
Rice straw and pig manure	Li et al. (2015b)
Biodiesel waste glycerin and municipal wastewater sludge	Razaviarani and Buchanan (2015)
Olive mill wastewater and liquid poultry manure	Khoufi et al. (2015)
Sewage sludge and sugar beet pulp	Montanes et al. (2015)
Pig manure and algae	Astals et al. (2015)
Forage radish and dairy manure	Belle et al. (2015)

preferably by a process favoring homo fermentative lactobacilli to minimize carbon loss (Gassen 2005). Biogas formation from plant fibres is generally a three-stage process involving a different set of anaerobic and facultative anaerobic microorganisms in each stage:

A. Hydrolysis of biomolecules
B. Acetogenesis: the production of acetic acid and carbon dioxide.
C. Methanogenesis with up to 70% (v/v) CH_4 and 30% CO_2 and the by-products NH_3 and H_2S by slow-growing archaea, which are sensitive to acidification, ammonia accumulation, low amounts of oxygen and other factors.

The bacterial community engaged in these three stages may be similar to those in cows rumen (Einspanier et al. 2004) or wastewater treatment plants (Ariesyady et al. 2007).

Further development of biogas technology is expected to increase production efficiency. Presently, only up to a maximum of about 70% of the organic matter in biomass is converted to CH_4 and CO_2. In order for this to increase, the hydrolysis stage must be enhanced. The separation of the processes for hydrolysis and for acetogenesis/methanogenesis allows for the application of different optimized conditions in the two stages, such as pH and temperature adjustment. Aside from the traditional mesophilic processes, thermophilic processes are being used more frequently to speed up the reactions and especially to optimize biomass hydrolysis. However, whereas in many industrial biogas plants the separation of the hydrolysis stage has already been carried out, most agricultural biogas plants use the single-stage technology.

Dried and desulfurized biogas is usually fuelled without CO_2 separation into stationary block heat and power plants connected to the biogas plants. Utilization of the excess heat is rarely possible because farms are usually located far away from residential or industrial areas where it could be used for domestic heating or manufacturing processes. This is not a problem if biogas is compressed like compressed natural gas stored in high-pressure cylinders and used in the engines of urban co-generation plants. In addition, direct use in car combustion engines is possible.

1.4.2 Composition of Biogas

Methane and carbon dioxide are major constituents of biogas with several impurities. Its general characteristics have been listed in Table 1.7 already. Biogas containing methane ratio above 45% is flammable. Influence of biogas components on its quality has been shown in Table 1.9. It gives general idea about usual gas constituents and their impact on the burning capacity of the biogas.

Table 1.9 Typical components and impurities in biogas

Component	Content (volume)	Effect	References
CO_2	25–50%	Lowers the calorific value Increases the methane number and the anti-knock properties of engines Causes corrosion (low concentrated carbon acid). If the gas is wet	Rasi et al. (2007)
H_2S	0–0.5%	Damages alkali fuel cells Corrosive effect in equipment and piping systems (stress corrosion); many manufacturers of engines therefore set an upper limit of 0.05 by vol.%; SO_2 emissions after burners or H_2S emissions with imperfect combustion—upper limit 0.1 by vol.% Spoils catalysts	Soroushian et al. (2006)
NH_3	0–0.05%	NO_x emissions after burners damage fuel cells Increases the anti-knock properties of engines	Burch and Southward (2000)
Water vapors	1–5%	Causes corrosion of equipment and piping systems Condensates damage instruments and plants Risk of freezing of piping systems and nozzles	Kapdi et al. (2005)
Siloxanes	0–50 mg/m^3	Act like an abrasive and damages engines	Dewil et al. (2006)

1.4.2.1 Methane and Carbon Dioxide

Methane to carbon dioxide ratio in the biogas can be managed to some extent. Following factors majorly effect:

- The presence of material rich in long chain hydrocarbon compounds having fat, can enhance the biogas quality
- More liquid in the bioreactor can decrease the CO_2 concentration in biogas as the water keep dissolving the CO_2, so reducing in the gas phase.
- Higher temperature during process of fermentation leads to low concentration of CO_2 dissolved in water.
- More CO_2 is dissolved in water at higher pressures.

The content of hydrogen sulfide in biogas mostly depends on the process and the type of waste used. Without a desulfurizing step, the concentration of H_2S would often exceeds 0.2% by volume.

1.4.3 Paybacks from Biogas Production

Biogas is a renewable energy source and has many applications. Several profits have to be derived from the conversion of various substrates in a biogas plant:

- In many countries, governments subsidize the erection of biogas plants to give the farmers an additional fuel source.
- Production of biogas from agricultural crops may maintain the structure of the landscape.
- Left over agricultural residues that are no more wanted are frequently prone to decomposition, but bioenergy can be generated.
- Landfill area can be minimized with the protection of the groundwater.
- Throwing away expenses of organic materials are reduced.
- Using plants as co-substrates increase the chances for recycling of the mineral fertilizer.
- CO_2 neutral production of energy is achieved.

1.5 Bioethanol

As the world population is increasing, the typical calorie consumption is on rise; thus enhancing the pressure on production from rare arable land but simultaneously, the energy requirement by developing nations is also increasing and the additional fuel most likely will be demanded from alternative renewable sources such as biofuel (Graham-Rowe 2011; Dutta et al. 2014).

In first generation ethanol production processes, readily available sugars or starch are utilized. The ethanol produced is readily used in today's engines. During the process, CO_2 is extracted from carbohydrates, which have a C/H/O ratio of 1:2:1 (Arshad et al. 2017). Ethanol, with its high (C + H) to O ratio, retains most of the original energy content. Because cell can produce much less energy from this anaerobic reaction than from oxidative respiration, it has to consume about ten times the amount of substrate to gain the same amount of energy; 2–3 ATP compared to 26–38 ATP in oxidative respiration, depending on the organism. This higher turnover of substrate is an advantage for biotechnology. This anaerobic fermentation also helps to avoid energy intensive aeration during industrial production (Antoni et al. 2007).

1.5.1 Ethanol as Fuel in History

Since the humanity exists, the biofuel are in use over the history. Mankind had relied on renewable energy resources like wood, windmills, water wheels and animals such as horses and oxen. Exploration of new energy resources was a major driving force behind technological revolution. In the start of nineteenth century, alcohols were over and over again reported as biofuel with the invention of ignition engines using biofuel. Nikolaus August Otto used ethanol for his spark ignition engine in the 1860s. Henry Ford also marketed his Model T, totally operating on 100% ethanol (Kovarik 1998). Ethanol production was widely abolished due to the unbeatably low price of gasoline in the USA. Ethanol as a fuel was revived in the 1970s in Brazil where one of the largest bioethanol industries is located today. Like modern crude oil refinery, the bio-industry for biofuel has a dual purpose in the economy, as it is used as a supply of energy as well as basic chemicals (Zaborsky 1982). The upcoming "bio refinery" revitalizes the old tradition of a careful thrifty economy and intends to make use of all energy and carbon stored in biomass, feeding byproducts into secondary conversion process or refining them as fuel.

1.5.2 Ethanol Fermentation

Glycolysis is the series of reactions taking place entirely in the cytosol, is the process of intracellular transformation of hexoses (glucose and fructose) into pyruvate with the formation of ATP and NADH (Zamora 2009). In the beginning, sugars are shifted inside the cell through facilitated diffusion (Weusthuis et al. 1994). Yeast cells keep many glucose transporters such as Gal2, Hxt1, Hxt2, Hxt3, Hxt4, Hxt6 and Hxt7 (Maier et al. 2002). Firstly, the glucose is converted to fructose 1,6-biphosphate. The reaction requires 2 ATP molecules, comprising three steps (Ratledge 1991; Bellou et al. 2014). In the second phase, glyceraldehyde-3-phosphate and dihydroxyacetone phosphate are made (Aggelis 2007). Then, glyceraldehyde-3-phosphate is transferred to 1,3-biphosphoglycerate.

The reaction catalyzed by glyceraldehyde 3-phosphate dehydrogenase, involves the synthesis of one mole of NADH. Afterward,1,3-Biphospho Glycerate is transferred into 3-phosphoglycerate, reaction catalyzed by phosphoglycerate kinase, with simultaneous release of one mole of ATP. In the end 3-phosphoglycerate is converted into pyruvate which is the final product of glycolysis, with immediate formation of another mole of ATP (Aggelis 2007; Festel 2008; Arshad et al. 2014b). So in this way, one mole of glucose in glycolysis creates two moles of pyruvic acid and NADH with four moles of ATP. As two moles of ATP are consumed to activate a mole of hexose molecule, balanced energy gain in glycolysis for the cell is remains only two ATP per hexose metabolized. Now pyruvate formed through glycolysis can be utilized by yeasts in different metabolic pathways.

Obviously, the microbes have to regenerate NAD^+ from the NADH to restore the oxidation-reduction potential of the cell and done through fermentation or respiration. Here the common trunk of glycolysis ends. Further, to proceed through alcoholic fermentation, glycerol-pyruvic fermentation or respiration depends upon various conditions (Rib′ereau-Gayon et al. 2006; Zamora 2009). In anaerobic conditions, the reducing power of NADH produced through glycolysis must be transferred to an electron acceptor to regenerate NAD^+ consumed by glycolysis. The process is called alcoholic fermentation and occurs in the cytoplasm, where acetaldehyde accepts the electrons (Ratledge 1991).

In addition to glycolysis, two additional enzymatic reactions occur in alcoholic fermentation. Pyruvate decarboxylase performs decarboxylation of pyruvate into acetaldehyde, using cofactors thiamine pyrophosphate and magnesium. In the end acetaldehyde is reduced into ethanol recycling NADH to NAD^+ by the alcohol dehydrogenase enzyme using zinc as cofactor. The final products of alcoholic fermentation, carbon dioxide and ethanol, are simply diffused out of the cell (Arshad et al. 2011).

1.5.3 Substrates Utilized for Bioethanol Production

Bioethanol fermentation is considerably the largest scale microbial process. Regardless of the simple or complex substrates utilized as microbial carbon sources acquiescent for conversion to ethanol, all types of substrates firstly result in the formation of hexoses, pentoses or glycerol (after the enzymatic, physical, chemical or mechanical pretreatment) that will be fermented by the relevant microorganisms in order to be converted into bioethanol. Major types of feedstock used in fuel ethanol production are presented in Table 1.10. In industrial ethanol production sugars present in sugar cane molasses or from enzymatically hydrolysed starch (from corn or other grains) and batch fermentation with yeast *Saccharomyces cerevisiae* is employed to produce ethanol.

Byproducts of the process include CO_2 with low amounts of methanol, glycerol, higher alcohols and acetic acid (Arshad et al. 2008). Ethanol does not need to be rectified to high purity if it is to be used as a fuel. Alcoholic fermentation process of sugars to ethanol has been well progressed in recent years. Alcoholic fermentation biochemistry includes substrate degradation pathways (glycolysis, alcoholic fermentation, glycerol-pyruvic fermentation and respiration for the case of the utilization of hexoses, xylose catabolic pathways for the case of utilization of pentoses and glycerol assimilation and glycolysis for the case of glycerol-converting

Table 1.10 Major feedstock used for fuel ethanol production

Feedstock	References
Sugar cane juice	Moreira and Goldemberg (1999)
Caasava	Agrocadenas (2006)
Sugarbeet	Poitrat (1999)
Wheat	Agrocadenas (2006)
Corn	Shapouri et al. (2003)
Sugarcane bagasse	Moreira (2000)
Corn stover	Kim and Dale (2004)
Wheat straw	Kim and Dale (2004)
Biomass	Berg (2001)

microorganisms) and regulation between fermentation and respiration (Pasteur effect, Crabtree effect, Kluyver effect and Custers effect).

Inhibitor sensitivity, product tolerance, ethanol yield and specific ethanol productivity have been improved in modern industrial strains to the degree that up to 20% (v/v) of ethanol are produced in present-day industrial yeast fermentation vessels from starch derived glucose. Substrates used for bioethanol production can be categorized into three major types:

1.5.3.1 Feedstock Containing Sucrose

Major feed stocks containing sucrose are sugarcane and sugar beet. Approximately 70 and 110 L/ton ethanol is produced from sugar cane and sugar beet respectively. Brazil alone produces 40% of world sugarcane. Sugar beet is the major feedstock for bioethanol production in European countries.

1.5.3.2 Starch Containing Feedstock

In Europe and North America, ethanol is majorly produced from starch containing feedstock such as corn, wheat and barley. In starch, D-glucose is linked through α-1,4 linkage with specific branches of 1-6 bonds. Conversion of starch into its monomer glucose is must require for ethanol fermentation. Corn is fermented into ethanol, starting by either dry- or wet-milling.

1.5.3.3 Lignocellulose Biomass

Present corn-based ethanol production may not be socio economically sustainable due to its impact on agricultural land usage and water shortage. The potential alternative substrate is lignocellulose biomass for ethanol production. Advantages and disadvantages about different methods for pretreatment of lignocellulose materials have been presented in Table 1.11. The lignocellulose biomass includes wood, straw, grasses, crop residues and other agricultural wastes, available in much higher quantities as compare to starch and sucrose containing substrates. Glucose yield is although much lower in cellulosic biomass as compared to sugar or starch crops, but easily accessible vast mass makes it better option for fuel ethanol production. Estimated potential of bioethanol production from agricultural residues is about 491 billion L/year.

Important difference between sugar and lignocellulosic material is the readily availability of substrate for fermentation. The technical process using lignocellulosic hydrolysates (Gray et al. 2006) is going to be better day by day. However, as the enzymatic hydrolysis reaction of cellulose is about two orders of magnitude slower than the average ethanol fermentation rate with yeast, there is a theoretical gap in simultaneous scarification of cellulosic biomass and ethanol fermentation

Table 1.11 Advantages and disadvantages about different methods for pretreatment of lignocellulose materials (Harmsen et al. 2010)

Pretreatment process	Benefits	Drawbacks
Biological	Lignin and hemicellulose can be decomposed easily Least energy needed	Hydrolysis reaction is much sluggish
Milling	Crystal structure of cellulose can be relaxed	Additional power required
Steam explosion	Lignin can be converted to its components easily Better release of glucose molecules	Discharge of toxic compounds Partial degradation of hemicellulose
CO_2 explosion	Much surface area exposed Better in terms of cost No toxic compounds generation	High pressure needed
Wet oxidation	Better abstraction of lignin Reduced formation of inhibitors	High cost
Organosolv	Efficient decomposition of lignin and hemicellulose	Drainage can cause environmental issues
Diluted acid	Lesser corrosion issues as compared to concentrated acid	Byproducts are formed
Concentrated acid	Much better glucose production	Acid recovery is essential

Table 1.12 Various fermentation processes for ethanol production

Mode of fermentation	Ethanol (g/L) in fermentation broth	Productivity, g/(L h)	Maximum yield (%)	References
Batch	80–100	1–3	85–90	Claassen et al. (1999)
Fed batch	53.7–98.1	9–31	81	Echegaray et al. (2000)
Repeated batch	89.3–92	2.7–5.25	80.5	Hojo et al. (1999)
Continuous	70–80	7–8	94.5	Costa et al. (2001)

(SSF). This must be addressed if total biomass is to be fermented, not only glucose syrups, for example from starch. Table 1.12 shows various fermentation processes for ethanol production. The fermentation of pentose sugar with industrial yeast strains is a difficult task and still under development (Hahn-Hagerdahl et al. 2007), although some pilot plants are already running.

Biological ethanol fermentation from molasses and starch is basically a mature technology. The utilization of non-food substrates such as cellulose-containing waste material is in the pilot stage. The hydrolysis of cellulosic material by the cellulase enzymes is very slow due to less porosity, crystal structure and presence of lignin and hemicellulose (Karim et al. 2017). More over the process also needs utilization of C_5 sugars to be economical. There are several approaches for the pretreatment of such biomass to release the required sugars. Physical, chemical and enzymatic treatments are availed. Availability of feedstock is not uniform throughout the year and also varies from region to region. Production cost of bioethanol is depended on the price of feedstock (Ray et al. 2017).

1.5.4 Ethanol Purification

Many yeast strains and their varieties are employed for industrial ethanol production. Certain strains lead over others, in specific rate of fermentation, better yield, efficient sugar utilization and higher tolerance of ethanol (Choudhary et al. 2017). But the byproducts are unavoidable in each and every strain. Formation of byproduct also depends on the purity of the substrate used. Acetaldehyde is one of the major byproduct of this process. Some higher alcohols as isoamyle alcohol are also produced as byproducts.

1.5.5 Ethanol as Fuel

Unlike petroleum, ethanol comes from renewable resources. It keeps cleaner burning characteristics (Prasad et al. 2007) as compared to gasoline; thus produces less greenhouse gases (McMillan 1997; Alzate and Toro 2006; Marchetti et al. 2007). Use of agro-industrial residues as raw material for ethanol fermentation, not only provides alternative substrates but also reduces carbon dioxide emissions with solution of their disposal problems. As ethanol is a biodegradable and comparatively highly soluble in water, has low toxicity risk. In case of any large spilling, far less danger for the environment than those associated with conventional oils (McMillan 1997). The potential of bioethanol production in totally non-aseptic environment (Roukas 1995; Kopsahelis et al. 2012; Sarris et al. 2013; Sarris et al. 2014) makes the process easier to apply at industrial scale. The use of Ethanol in place of petroleum could, provided that a renewable energy resource was used to produce crops required to obtain ethanol and to distil fermented ethanol.

Ethanol is the compound of carbon and hydrogen atoms, with a hydroxyl group have chemical formula C_2H_5OH, also known as ethyl alcohol or hydroxyl ethane. Its molecule is small and light, as compared to most gasoline components. The electrochemistry of the ethanol molecule is slightly exceptional being polar at one end and non-polar at the other. It participates in hydrogen linkage with other ethanol molecules or other polar substances due to presence of hydroxyl group. The polar end help ethanol to be miscible in water or other polar compounds and the non-polar end is advantageous in mixing with non-polar substances, such as gasoline. Generally ethanol is produced in two forms: anhydrous, keeping water content less than 1%, or hydrous, having water content up to 10%. Purity of ethanol above 96% cannot be achieved through conventional distillation. To convert hydrous ethanol into anhydrous, separate technique as azeotropic distillation using 20–25% extra energy is used.

Generally ethanol in blended with gasoline in percentages from 5 to 85% (Kim and Dale 2006) and diesel for its use as fuel. Above half of the fuel ethanol used worldwide is blended with gasoline. Globally most prevalent blends are E_{85}, E_{20} and E_{10} (Festel 2008). Primary reason for using ethanol as an additive to gasoline is reduction in CO_2 emissions. Ethanol addition also raises the octane number of the fuel blend thus it can replace more costly octane-boosting components such as alkylate. Ethanol keeps oxygen, so gasoline burns more cleanly and reduces the amount of harmful emissions of carbon monoxide (CO), particulates and unburned gasoline components. Ethanol can be used in the trans-esterification of vegetable oils for the production of fatty acid ethyl esters (Marchetti et al. 2007).

1.6 Conclusion

Owing to rapid growth of biofuel production in last decades biofuel are fulfilling almost 3% of transport fuel needs worldwide. Such a rapid increase has given rise to many concerns. World has faced increased in food prices worldwide and an alarm to food security. Major advantage of biofuel in climate change mitigation has been also facing questions. As the per energy equivalence, factory gate prices of ethanol and biodiesel were almost higher as compared to refinery gate prices of fossil based gasoline and diesel. Production costs of biofuel needs substantial reduction to make these products completive. The technological breakthroughs can do the best in future. Present production ways and techniques has been comprehensively discussed above.

References

Abdeshahian P, Lim JS, Ho WS, Hashim H, Lee CT. Potential of biogas production from farm animal waste in Malaysia. Renew Sust Energ Rev. 2016;60:714–23.

Abdmouleh Z, Alammari RA, Gastli A. Review of policies encouraging renewable energy integration & best practices. Renew Sust Energ Rev. 2015;45:249–62.

Aboudi K, Alvarez-Gallego CJ, Romero-Garcia LI. Semi-continuous anaerobic co-digestion of sugar beet byproduct and pig manure: effect of the organic loading rate (OLR) on process performance. Bioresour Technol. 2015;194:283–90.

Aggelis G. Microbiology and microbial technology, A. Athens, Greece: Stamoulis Publishers; 2007.

Agrocadenas. Segundo informe de coyunturamaı´z 2006. Observatorio Agrocadenas Colombia, Ministry of Agricultural and Rural Development. http://www.agrocadenas.gov.co/home.htm. Accessed Feb 2007.

Ahmad AL, Yasin NHM, Derek CJC, Lim JK. Microalgae as a sustainable energy source for biodiesel production: a review. Renew Sust Energ Rev. 2011;15:584–93.

Alam M, Hasan M. Feasibility study of biogas energy in Bangladesh (Doctoral dissertation) Daffodil International University; 2017.

Alonso DM, Bond JQ, Dumesic JA. Catalytic conversion of biomass to biofuels. Green Chem. 2010;12:1493–513.

Alzate CAC, Toro OJS. Energy consumption analysis of integrated flow sheets for production of fuel ethanol from lignocellulosic biomass. Energy. 2006;31:2111–23.

Amin S. Review on biofuel oil and gas production processes from microalgae. Energ convers manage. 2009;50:1834–40.

Antoni D, Zverlov VV, Schwarz WH. Biofuels from microbes. Appl Microbiol Biotechnol. 2007;77(1):23–35.

Ariesyady HD, Ito T, Okabe S. Functional bacterial and archaeal community structures of major trophic groups in a fullscale anaerobic sludge digester. Water Res. 2007;41:1554–68.

Ariunbaatar J, Panico A, Esposito G, Pirozzi F, Lens PNL. Pretreatment methods to enhance anaerobic digestion of organic solid waste. Appl Energ. 2014;123:143–56.

Arshad M. Bioethanol: A sustainable and environment friendly solution for Pakistan. A Scientific J. COMSATS–Sci. Vision. 2010;16–7.

Arshad M, Ahmed S. Cogeneration through bagasse: a renewable strategy to meet the future energy needs. Renew Sust Energ Rev. 2016;54:732–7.

Arshad M, Khan ZM, Shah FA, Rajoka MI. Optimization of process variables for minimization of byproduct formation during fermentation of blackstrap molasses to ethanol at industrial scale. Lett Appl Microbiol. 2008;47:410–4.

Arshad M, Zia MA, Asghar M, Bhatti H. Improving bio-ethanol yield: Using virginiamycin and sodium flouride at a Pakistani distillery. Afr J Biotechnol. 2011;10:11071.

Arshad M, Adil M, Sikandar A, Hussain T. Exploitation of meat industry by-products for biodiesel production: Pakistan's perspective. Pakistan J Life Soc Sci. 2014a;12:120–5.

Arshad M, Ahmed S, Zia MA, Rajoka MI. Kinetics and thermodynamics of ethanol production by *Saccharomyces cerevisiae* MLD10 using molasses. Appl Biochem Biotechnol. 2014b;172:2455–64.

Arshad M, Hussain T, Iqbal M, Abbas M. Enhanced ethanol production at commercial scale from molasses using high gravity technology by mutant *S. cerevisiae*. Brazilian J Microbiol. 2017. doi:10.1016/j.bjm.2017.02.003.

Astals S, Musenze RS, Bai X, Tannock S, Tait S, Pratt S, et al. Anaerobic co-digestion of pig manure and algae: impact of intracellular algal products recovery on co-digestion performance. Bioresour Technol. 2015;181:97–104.

Atabani AE, Silitonga AS, Badruddin IA, Mahlia TMI, Masjuki HH, Mekhilef S. A comprehensive review on biodiesel as an alternative energy resource and its characteristics. Renew Sust Energ Rev. 2012;16:2070–93.

Balat M. Potential alternatives to edible oils for biodiesel production-A review of current work. Energ Convers Manage. 2011a;52:1479–92.

Balat M. Production of bioethanol from lignocellulosic materials via the biochemical pathway: a review. Energ Convers Manag. 2011b;52:858–75.

Balat M, Balat H. Progress in biodiesel processing. Appl Energ. 2010;87:1815–35.

Balat M, Balat M, Kırtay E, Balat H. Main routes for the thermo-conversion of biomass into fuels and chemicals. Part 1: Pyrolysis systems. Energ Convers Manage. 2009;50:3147–57.

Banković-Ilić IB, Stamenković OS, Veljković VB. Biodiesel production from non-edible plant oils. Renew Sust Energ Rev. 2012;16:3621–47.

Barney BM. Metabolic engineering: the sweet smell of biosynthesis. Nat Chem Biol. 2014;10:246–7.

Bart JCJ, Palmeri N, Cavallaro S. Biodiesel science and technology: from soil to oil. Woodhead Publishing Limited; 2010.

Basha SA, Gopal KR, Jebaraj S. A review on biodiesel production, combustion, emissions and performance. Renew Sust Energ Rev. 2009;13:1628–34.

Basumatary S. Yellow Oleander (*Thevetia peruviana*) seed oil biodiesel as an alternative and renewable fuel for diesel engines: a review. Int J Chem Tech Res. 2015;7:2823.

Battista F, Fino D, Erriquens F, Mancini G, Ruggeri B. Scaled-up experimental biogas production from two agro-food waste mixtures having high inhibitory compound concentrations. Renew Energ. 2015;81:71–7.

Belle AJ, Lansing S, Mulbry W, Weil RR. Anaerobic co-digestion of forage radish and dairy manure in complete mix digesters. Bioresour Technol. 2015;178:230–7.

Bellou S, Baeshen MN, Elazzazy AM, Aggeli D, Sayegh F, Aggelis G. Microalgal lipids biochemistry and biotechnological perspectives. Biotechnol Adv. 2014;32:1476–93.

Bentley RW. Introduction. In: Introduction to peak oil. Springer International Publishing; 2016. P. 1–8.

Berchmans HJ, Hirata S. Biodiesel production from crude *Jatropha curcas* L. seed oil with a high content of free fatty acids. Bioresour Technol. 2008;99:1716–21.

Berg C. World fuel ethanol. Analysis and outlook. F.O. Licht. http://www.agra-europe.co.uk/FOLstudies/FOL-Spec04.html (2001). Accessed March 2004.

Bergthorson JM, Thomson MJ. A review of the combustion and emissions properties of advanced transportation biofuels and their impact on existing and future engines. Renew Sust Energ Rev. 2015;42:1393–417.

Bhola V, Swalaha F, Kumar RR, Singh M, Bux F. Overview of the potential of microalgae for CO_2 sequestration. Int J Environ Sci Technol. 2014;11:2103–18.

Bilgen S. Structure and environmental impact of global energy consumption. Renew Sust Energ Rev. 2014;38:890–902.

Borras SM, McMichael P, Scoones I. 2010 The politics of biofuels, land and agrarian change: a special issue on biofuels. J. Peasant Stud. 2010;37:575–92.

Borse P, Sheth A. Technological and commercial update for first and second generation ethanol production in India. In: Sustainable biofuels development in India. Springer International Publishing; 2017. PP. 279–97.

Brennan L, Owende P. Biofuels from microalgae-a review of technologies for production, processing, and extractions of biofuels and co-products. Renew Sust Energ Rev. 2010;14: 557–77.

Brown TM, Duan P, Savage PE. Hydrothermal liquefaction and gasification of Nannochloropsis sp. Energ Fuels. 2010;24:3639–46.

Burch R, Southward BWL. A novel application of trapping catalysts for the selective low-temperature oxidation of NH_3 to N_2 in simulated biogas. J Catal. 2000;195:217–26.

Calero J, Luna D, Sancho ED, Luna C, et al. An overview on glycerol-free processes for the production of renewable liquid biofuels, applicable in diesel engines. Renew Sust Energ Rev. 2015;42:1437–52.

Canakci M, Ozsezen AN, Arcaklioglu E, Erdil A. Prediction of performance and exhaust emissions of a diesel engine fueled with biodiesel produced from waste frying palm oil. Expert Syst Appl. 2009;36:9268–80.

Carriquiry MA, Du X, Timilsina GR. Second generation biofuels: economics and policies. Energ Policy. 2011;39:4222–34.

Castañeda-Ayarza JA, Cortez LAB. Final and B molasses for fuel ethanol production and some market implications. Renew Sust Energ Rev. 2016.

Chapagain BP, Yehoshua Y, Wiesman Z. Desert date (*Balanites aegyptiaca*) as an arid lands sustainable bioresour for biodiesel. Bioresour Technol. 2009;100:1221–6.

Chavez-Baeza C, Sheinbaum-Pardo C. Sustainable passenger road transport scenarios to reduce fuel consumption, air pollutants and GHG (greenhouse gas) emissions in the Mexico City Metropolitan Area. Energy. 2014;66:624–34.

Chhetri AB, Tango MS, Budge SM, Watts KC, Islam MR. Non-edible plant oils as new sources for biodiesel production. Int J Mol Sci. 2008;9:169–80.

Chisti Y. Biodiesel from microalgae. Biotechnol Adv. 2007;25(3):294–306.

Choi SP, Nguyen MT, Sim SJ. Enzymatic pretreatment of *Chlamydomonas reinhardtii* biomass for ethanol production. Bioresour Technol. 2010;101:5330–6.

Choudhary J, Singh S, Nain L. Bioprospecting thermo tolerant ethanologenic yeasts for simultaneous saccharification and fermentation from diverse environments. J Biosci Bioeng. 2017;123:342–6.

Claassen PAM, van Lier JB, Lo´pez Contreras AM, van Niel EWJ, Sijtsma L, Stams AJM, et al. Utilisation of biomass for the supply of energy carriers. Appl Microbiol Biotechnol. 1999;52:741–55.

Costa AC, Atala DIP, Maugeri F, Maciel R. Factorial design and simulation for the optimization and determination of control structures for an extractive alcoholic fermentation. Process Biochem. 2001;37:125–37.

Cremonez PA, Feroldi M, Nadaleti WC, deRossi E, Feiden A, DeCamargo MP, et al. Biodiesel production in Brazil: current scenario and perspectives. Renew Sust Energ Rev. 2015;42:415–28.

Creutzig F, Ravindranath NH, Berndes G, Bolwig S, Bright R, Cherubini F, Chum H, Corbera E, Delucchi M, Faaij A, Fargione J. Bioenergy and climate change mitigation: an assessment. Gcb Bioenerg. 2015;7:916–44.

Dai W, Word DP, Hahn J. Modeling, dynamic optimization of fuel-grade ethanol fermentation using fed-batch process. Control Eng Practice. 2014;22:231–41.

Dale BE, Anderson JE, Brown RC, Csonka S, Dale VH, Herwick G, et al. Take a closer look: biofuels can support environmental, economic and social goals. Environ Sci Technol. 2014;48:7200–3.

Danielsen F, Beukema H, Burgess ND, Parish F, Bruhl CA, Donald PF, et al. Biofuel plantations on forested lands: double jeopardy for biodiversity and climate. Conserv Biol. 2009;23:348–58.

De Francisci D, Kougias P, Treu L, Campanaro S, Irini A. Microbial diversity and dynamicity of biogas reactors fed with different substrates. In 2nd international conference on biogas microbiology; 2014.

De Kam MJ, Morey RV, Tiffany DG. Biomass integrated gasification combined cycle for heat and power at ethanol plants. Energ Convers Manage. 2009;50:1682–90.

deAlegría IM, Basañez A, deBasurto PD, Fernández-Sainz A. Spain′ s fulfillment of its Kyoto commitments and its fundamental greenhouse gas (GHG) emission reduction drivers. Renew Sust Energ Rev. 2016;59:858–67.

Demirbas A, Demirbas MF. Importance of algae oil as a source of biodiesel. Energ Convers Manage. 2011;52:163–70.

Demirbas A. Relationships derived from physical properties of vegetable oil and biodiesel fuels. Fuel. 2008;87:1743–8.

Demirbas A. Biodiesel from waste cooking oil via base-catalytic and supercritical methanol transesterification. Energ Convers Manage. 2009a;50:923–7.

Demirbas A. Diesel-like fuel from tallow by pyrolysis and supercritical water liquefaction. Energ Source A. 2009b;31:824–30.

Demirbas A. Progress and recent trends in biodiesel fuels. Energ Convers Manage. 2009c;50:14–34.

Demirbas A. Biofuels securing the planet's future energy needs. Energ Convers Manage. 2009d;50:2239–49.

Deng X, Fang Z, Liu YH, Yu CL. Production of biodiesel from Jatropha oil catalyzed by nanosized solid basic catalyst. Energy. 2011;36:777–84.

Dewil R, Appels L, Baeyens J. Energy use of biogas hampered by the presence of siloxanes. Energ Convers Manage. 2006;47:1711–22.

Dias JM, Alvim-Ferraz MC Almeida MF. Mixtures of vegetable oils and animal fat for biodiesel production: influence on product composition and quality. Energ Fuels. 2008;22:3889–93.

Diouf J. Biofuels a disaster for world food. See http://eucoherence.org/renderer.do/,2007. ClearState/false/menuld/22735/return.

Divya D, Gopinath LR, Christy PM. A review on current aspects and diverse prospects for enhancing biogas production in sustainable means. Renew Sust Energ Rev. 2015;42:690–9.

Duffield JA, Johansson R, Meyer S. US ethanol an examination of policy, production, use, distribution, and market interactions; 2015.

Dutta K, Daverey A, Jih-Gaw LJG. Evolution retrospective for alternative fuels: first to fourth generation. Renew Energ. 2014;69:114–22.

Dwivedi P, Wang W, Hudiburg T, Jaiswal D, Parton W, Long S, et al. Cost of abating greenhouse gas emissions with cellulosic ethanol. Environ Sci Technol. 2015;49:2512–22.

Echegaray O, Carvalho J, Fernandes A, Sato S, Aquarone E, Vitolo M. Fed-batch culture of *Saccharomyces cerevisiae* in sugarcane blackstrap molasses: invertase activity of intact cells in ethanol fermentation. Biomass Bioenerg. 2000;19:39–50.

Einspanier R, Lutz B, Rief S, Berezina O, Zverlov V, Schwarz WH, Mayer J. Tracing residual recombinant feed molecules during digestion and rumen bacterial diversity in cattle fed transgene maize. Eur Food Res Technol. 2004;218:269–73.

Elbehri A, Segerstedt A, Liu P. Biofuels and the sustainability challenge: a global assessment of sustainability issues, trends and policies for biofuels and related feedstocks. Food and Agriculture Organization of the United Nations (FAO); 2013.

El-Mashad HM, Zhang R. Biogas production from co-digestion of dairy manure and food waste. Bioresour Technol. 2010;101:4021–8.

Emanuel E, Gomes C. An assessment of mechanisms to improve energy efficiency in the transport sector in Grenada, Saint Lucia and Saint Vincent and the Grenadines 2014.

Ertas M, Alma MH. Slow pyrolysis of chinaberry (*Meliaazedarach L.*) seeds: Part I. The influence of pyrolysis parameters on the product yields. Ener Edu Sci Technol Part A-Energ Sci Res. 2011;26:143–54.

Eryilmaz T, Yesilyurt MK, Cesur C, Gokdogan O. Biodiesel production potential from oil seeds in Turkey. Renew Sust Energ Rev. 2016;58:842–51.

Escobar JC, Lora ES, Venturini OJ, Yáñez EE, et al. Biofuels: environment, technology and food security. Renew Sust Energ Rev. 2009;13:1275–87.

Esteves IA, Lopes MS, Nunes PM, Mota JP. Adsorption of natural gas and biogas components on activated carbon. Sep Purif Technol. 2008;62:281–96.

Falasca SL, Flores N, Lamas MC, Carballo SM, Anschau A. *Crambea byssinica*: an almost unknown crop with a promissory future to produce biodiesel in Argentina. Int J Hydrogen Energ. 2010;35:5808–12.

Fathya S, Assia K, Hamza M. Influence of inoculums/substrate ratios (ISRs) on the mesophilic anaerobic digestion of slaughterhouse waste in batch mode: process stability and biogas production. Energ Procedia. 2014;50:57–63.

Fernandes DM, Serqueira DS, Portela FM, Assunção RMN, Munoz RAA, Terrones MGH. Preparation and characterization of methylic and ethylic biodiesel from cottonseed oil and effect of tert-butylhydroquinone on its oxidative stability. Fuel. 2012;97:658–61.

Festel GW. Biofuels-economic aspects. Chem Eng Technol. 2008;31:715–20.

Fogel C. Constructing progressive climate change norms: the US in the early 2000s. The social construction of climate change, power, knowledge, norms, discourses. 2007;123–47.

Foong TM, Morganti KJ, Brear MJ, daSilva G, Yang Y, Dryer FL. The octane numbers of ethanol blended with gasoline and its surrogates. Fuel. 2014;115:727–39.

Friedlingstein P, Andrew RM, Rogelj J, Peters GP, Canadell JG, Knutti R, et al. Persistent growth of CO_2 emissions and implications for reaching climate targets. Nat Geosci. 2014;7:709–15.

Gassen HG. Biogasanlagen: Ein Beitrag zur umweltfreundlichen Energieversorgung. Biol unserer Zeit. 2005;35(6):384–92.

Géczi G, Borges CA, Ágoston C, Pusztai K, Ákos U Beszédes S. Examination of energy recovery of brewers'spent grain ii.-biological process. J Microbiol Biotechnol Food Sci. 2015;5:268.

Ghazali WNMW, Mamat R, Masjuki HH, Najafi G. Effects of biodiesel from different feedstocks on engine performance and emissions: a review. Renew Sust Energ Rev. 2015;51:585–602.

Gonzalez-Salazar MA, Venturini M, Poganietz WR, Finkenrath M, Kirsten T, Acevedo H, et al. Development of a technology roadmap for bioenergy exploitation including biofuels, waste-to-energy and power generation and CHP. Appl Energ. 2016;180:338–52.

Gouveia L, Oliveira AC. Microalgae as a raw material for biofuels production. J Indus Microbiol Biotechnol. 2009;36:269–74.

Graham-Rowe D. Beyond food versus fuel. Nature. 2011;474:S6–8.

Gray KA, Zhao L, Emptage M. Bioethanol. Curr Opin Chem Biol. 2006;10:141–6.

Gui MM, Lee KT, Bhatia S. Feasibility of edible oil vs. non-edible oil vs. waste edible oil as biodiesel feedstock. Energy. 2008;33:1646–53.

Hahn-Hägerdahl B, Karhumaa K, Fonseca C, Spencer-Martins I, Gorwa-Grauslund MF. Towards industrial pentose fermenting yeast strains. Appl Microbiol Biotechnol. 2007;74:937–53.

Haider MR, Zeshan YS, Malik RN, Visvanathan C. Effect of mixing ratio of food waste and rice husk co-digestion and substrate to inoculums ratio on biogas production. Bioresour Technol. 2015;190:451–7.

Haile H. Integrated volarization of spent coffee grounds to biofuels. Biofuel Res J. 2014;1:65–9.

Harmsen PFH, Huijgen W, Bermudez L, Bakker R. Literature review of physical and chemical pretreatment processes for lignocellulosic biomass (No. 1184). Wageningen UR Food and Biobased Research; 2010.

Haro P, Ollero P, Trippe F. Technoeconomic assessment of potential processes for bio-ethylene production. Fuel Process Technol. 2013;114:35–48.

Havukainen J, Uusitalo V, Niskanen A, Kapustina V, Horttanainen M. Evaluation of methods for estimating energy performance of biogas production. Renew Energ. 2014;66:232–40.

Headey D, Fan S. Anatomy of a crisis, the causes and consequences of surging food prices. Agric Econ. 2008;39:375–91.

Ho DP, Ngo HH, Guo W. A mini review on renewable sources for biofuel. Bioresour Technol. 2014;169:742–9.

Hoekman SK, Robbins C. Review of the effects of biodiesel on NO_x emissions. Fuel Process Technol. 2012;96:237–49.

Hoekman SK, Broch A, Robbins C, Ceniceros E, et al. Review of biodiesel composition, properties, and specifications. Renew Sust Energ Rev. 2012;16:143–69.

Hojo O, Hokka CO, SoutoMaior AM. Ethanol production by a flocculant yeast strain in a CSTR type fermentor with cell recycling. Appl Biochem Biotechnol. 1999;77–79:535–45.

Holmberg K, Erdemir A. Global impact of friction on energy consumption, economy and environment. FME Trans. 2015;43:181–5.

Höök M, Davidsson S, Johansson S, Tang X. Decline and depletion rates of oil production: a comprehensive investigation. Phil Trans R Soc A. 2014;372.20120448.

Hossain ABM, AlEissa MS. Biodiesel fuel production from palm, sunflower waste cooking oil and fish byproduct waste as renewable energy and environmental recycling process. 2016;10:1–9.

Hughes SR, Moser BR, Gibbons WR. Moving toward energy security and sustainability in 2050 by reconfiguring biofuel production. In: Convergence of food security, energy security and sustainable agriculture. Springer, Berlin Heidelberg; 2014:15–29.

Intergovernmental Panel on Climate Change (IPCC) 2014. Climate change 2014–impacts, adaptation and vulnerability: regional aspects. Cambridge University Press.

Janaun J, Ellis N. Perspectives on biodiesel as a sustainable fuel. Renew Sust Energ Rev. 2010;14:1312–20.

Janczak D, Kozlowski K, Zbytek Z, Cieslik M, Bugala A, Czekala W. Energetic efficiency of the vegetable waste used as substrate for biogas production. In MATEC Web of Conferences 2016;64. EDP Sciences.

Jatrofuels. From feedstock cultivation to full market integration. 2012 http://www.jatrofuels.com/161-0-Biofuels.html#Fuel%20characteristics%20and%20advantages.

Johari A, Nyakuma BB, Nor SHM, Mat R, Hashim H, Amad A, et al. The challenges and prospects of palm oil based biodiesel in Malaysia. Energy. 2015;81:255–61.

Jonker JGG, Van Der Hilst F, Junginger HM, Cavalett O, Chagas MF, Faaij APC. Outlook for ethanol production costs in Brazil up to 2030, for different biomass crops and industrial technologies. Appl Energ. 2015;147:593–610.

Kafle GK, Kim SH. Anaerobic treatment of apple waste with swine manure for biogas production: batch and continuous operation. Appl Energ. 2013;103:61–72.

Kafuku G, Mbarawa M. Biodiesel production from Croton megalocarpus oil and its process optimization. Fuel. 2010;89:2556–60.

Kalam MA, Masjuki HH. Biodiesel from palmoil-an analysis of its properties and potential. Biomass Bioenerg. 2002;23:471–9.

Kapdi SS, Vijay VK, Rajesh SK, Prasad R. Biogas scrubbing, compression and storage: perspective and prospectus in Indian context. Renew Energ. 2005;30:1195–202.

Karim Z, Afrin S, Husain Q, Danish R. Necessity of enzymatic hydrolysis for production and functionalization of nanocelluloses. Critic Rev Biotechnol. 2017;37:355–70.

Karmee SA, Chadha A. Preparation of biodiesel from crude oil of *Pongamia pinnata*. Bioresour Technol. 2005;96:1425–9.

Karwat DM, Eagle WE, Wooldridge MS. Are there ecological problems that technology cannot solve? Water scarcity and dams, climate change and biofuels. IJESJP. 2014;3:7–25.

Kern JD, Hise AM, Characklis GW, Gerlach R, Viamajala S, Gardner RD. Using life cycle assessment and techno-economic analysis in a real options framework to inform the design of algal biofuel production facilities. Bioresour Technol. 2017;225:418–28.

Khan SR. Viewing biofuel (ethanol) prospects in Pakistan through a sustainable development prism. SDPI Res News Bulletin. 2007;14.

Khoufi S, Louhichi A, Sayadi S. Optimization of anaerobic co-digestion of olive mill wastewater and liquid poultry manure in batch condition and semi-continuous jet-loop reactor. Bioresour Technol. 2015;182:67–74.

Kim S, Dale BE. Global potential bioethanol production from wasted crops and crop residues. Biomass Bioenerg. 2004;26:361–75.

Kim S, Dale B. Ethanol fuels: E10 or E85–life cycle perspectives. Int J Life Cycle Assess. 2006;11:117–21.

Knothe G, Krahl J, Van Gerpen J. eds. The biodiesel handbook. Elsevier. 2015.

Koberg M, Cohen M, Ben-Amotz A, Gedanken A. Bio-diesel production directly from the microalgae biomass of Nanno chloropsis by microwave and ultrasound radiation. Bioresour Technol. 2011;102:4265–9.

Koh MY, Ghazi MTI. A review of biodiesel production from Jatrophacurcas L. oil. Renew Sust Energ Rev. 2011;15:2240–51.

Kolbitsch P, Pfeifer C, Hofbauer H. Catalytic steam reforming of model biogas. Fuel. 2008;87:701–6.

Kopsahelis N, Bosnea L, Bekatorou A, Tzia C, Kanellaki M. Alcohol production from sterilized and non-sterilized molasses by *Saccharomyces cerevisiae* immobilized on brewer's spent grains in two types of continuous bioreactor systems. Biomass Bioenerg. 2012;45:87–94.

Kovarik B. Henry Ford, Charles Kettering, and the "fuel of the future". Automot Hist Rev. 1998;32:7–27.

Kristensen EF, Feng L, Møller HB. Storage and pretreatment of grass for from extensive lowland areas used in a biogas plant. In: International conference on agricultural engineering 2016.

Kumar A, Sharma S. Potential non-edible oil resources as biodiesel feedstock: an Indian perspective. Renew Sust Energ Rev. 2011;15:1791–800.

Kumar S, Shrestha P, Salam PA. A review of biofuel policies in the major biofuel producing countries of ASEAN: production, targets, policy drivers and impacts. Renew Sust Energ Rev. 2013;26:822–36.

Kusdiana D, Saka S. Effects of water on biodiesel fuel production by supercritical methanol treatment. Bioresour Technol. 2004;91:289–95.

Lee S, Speight JG, Loyalka SK. Handbook of alternative fuel technologies. CRC Press; 2014.

Leone TG, Olin ED, Anderson JE, Jung HH, Shelby MH, Stein RA. Effects of fuel octane rating and ethanol content on knock, fuel economy, and CO_2 for a turbocharged DI engine. SAE Int J Fuels Lubricants. 2014;7:9–28.

Leu JH. Biodiesel manufactured from waste cooking oil by alkali transerification reaction and its vehicle application. J Biobas Material Bioenerg. 2013;7:189–93.

Leung DYC, Wu X, Leung MKH. A review on biodiesel production using catalyzed transesterification. Appl Energ. 2010;87:1083–95.

Li D, Liu SC, Mi L, Li ZD, Yuan Y, Yan Z, et al. Effects of feedstock ratio and organic loading rate on the anaerobic mesophilic co-digestion of rice straw and cow manure. Bioresour Technol. 2015a;189:319–26.

Li D, Liu SC, Mi L, Li ZD, Yuan Y, Yan Z, et al. Effects of feedstock ratio and organic loading rate on the anaerobic mesophilic co-digestion of rice straw and pig manure. Bioresour Technol. 2015b;187:120–7.

Li LH, Sun YM, Yuan ZH, Kong XY, Wao Y, Yang L, et al. Effect of microalgae supplementation on the silage quality and anaerobic digestion performance of many flower silver grass. Bioresour Technol. 2015c;189:334–40.

Liu X, Piao X, Wang Y, Zhu S. Liquid–liquid equilibrium for systems of (fatty acid ethyl esters + ethanol + soybean oil and fatty acid ethyl esters + ethanol + glycerol). J Chem Eng Data. 2008;53:359–62.

Losordo Z, McBride J, Rooyen JV, Wenger K, Willies D, Froehlich A, et al. Cost competitive second-generation ethanol production from hemicellulose in a Brazilian sugarcane biorefinery. Biofuels, Bioprod Biorefin. 2016;10:589–602.

Luca C, Pilu R, Tambone F, Scaglia B, Adani FL. New energy crop giant cane *(Arundodonax L.)* can substitute traditional energy crops increasing biogas yield and reducing costs. Bioresour Technol. 2015;191:197–204.

Luterbacher JS, Rand JM, Alonso DM, Han J, Youngquist JT, Maravelias CT, et al. Nonenzymatic sugar production from biomass using biomass-derived γ-valerolactone. Science. 2014;343:277–80.

Lynd LR, Woods J. A new hope for Africa. Nature. 2011;474:S20–1.

Ma F, Hanna MA. Biodiesel production: a review. Bioresour Technol. 1999;70:1–15.

Maher KD, Bressler DC. Pyrolysis of triglyceride materials for the production of renewable fuels and chemicals. Bioresour Technol. 2007;98:2351–68.

Maier A, Volker B, Boles E, Fuhrmann GF. Characterisation of glucose transport in *Saccharomyces cerevisiae* with plasma membrane vesicles (counter transport) and intact cells (initial uptake) with single Hxt1, Hxt2, Hxt3, Hxt4, Hxt6, Hxt7 or Gal2 transporters. FEMS Yeast Res. 2002;2:539–50.

Majchrowicz E. ed. Biochemical pharmacology of ethanol. Springer Science & Business Media; 2013. P. 56.

Malakhova DV, Egorova MA, Prokudina LI, Netrusov AI, Tsavkelova EA. The biotransformation of brewer's spent grain into biogas by anaerobic microbial communities. World J Microbiol Biotechnol. 2015;31.

Mao C, Feng Y, Wang X, Ren G. Review on research achievements of biogas from anaerobic digestion. Renew Sust Energ Rev. 2015;2015(45):540–55.

Marchetti J, Miguel V, Errazu A. Possible methods for biodiesel production. Renew Sust Energ Rev. 2007;11:1300–11.

Maroun MR, La Rovere EL. Ethanol and food production by family small holdings in rural Brazil: economic and socio-environmental analysis of micro distilleries in the state of Rio Grande doSul. Biomass Bioenerg. 2014;63:140–55.

Martínez EJ, Gil MV, Fernandez C, Rosas JG, Gómez X. Anaerobic codigestion of sludge: addition of butcher's fat waste as a cosubstrate for increasing biogas production. PLoS ONE. 2016;11:e0153139.

Mata TM, Martins AA, Caetano NS. Microalgae for biodiesel production and other applications: a review. Renew Sust Energ Rev. 2010;14:217–32.

Mayer FD, Feris LA, Marcilio NR, Hoffmann R. Why small-scale fuel ethanol production in Brazil does not take off? Renew Sust Energ Rev. 2015;43:687–701.

McMillan JD. Bioethanol production: status and prospects. Renew Energ. 1997;10:295–302.

Meher LC, Sagar DV, Naik SN. Technical aspects of biodiesel production by transesterification-a review. Renew Sust Energ Rev. 2006;10:248–68.

Meng X, Yang J, Xu X, Zhang L, Nie Q, Xian M. Biodiesel production from oleaginous microorganisms. Renew Energ. 2009;34:1–5.
Metzger P, Largeau C. *Botryococcus braunii*: a rich source for hydrocarbons and related ether lipids. Appl Microbiol Biotechnol. 2005;66:486–96.
Mielenz J.R. Small-scale approaches for evaluating biomass bioconversion for fuels and chemicals. Bioenerg Biomass Biofuel. 2014;385.
Moeller L, Görsch K. Foam formation in full-scale biogas plants processing biogenic waste. Energ Sust Soc. 2015;5:1.
Mofijur M, Atabani AE, Masjuki HH, Kalam MA, Masum BM. A study on the effects of promising edible and non-edible biodiesel feedstocks on engine performance and emissions production: a comparative evaluation. Renew Sust Energ Rev. 2013;23:391–404.
Montanes R, Solera R, Perez M. Anaerobic co-digestion of sewage sludge and sugar beet pulp lixiviation in batch reactors: effect of temperature. Bioresour Technol. 2015;181:177–84.
Moreira JS. Sugarcane for energy-recent results and progress in Brazil. Energ Sust Develop. 2000;4:43–54.
Moreira J, Goldemberg J. The alcohol program. Energ Policy. 1999;27:229–45.
Murphy F, Devlin G, Deverell R, McDonnell K. Biofuel production in Ireland-an approach to 2020 targets with a focus on algal biomass. Energies. 2013;6:6391–412.
Murugesan A, Umarani C, Chinnusamy TR, Krishnan M, et al. Production and analysis of bio-diesel from non-edible oils-a review. Renew Sust Energ Rev. 2009;13:825–34.
Mutreja V, Singh S, Ali A. Biodiesel from mutton fat using KOH impregnated MgO as heterogeneous catalysts. Renew Energ. 2011;36:2253–8.
Nigam PS, Singh A. Production of liquid biofuels from renewable resources. Progress Energ Combust Sci. 2011;37:52–68.
Ohuchi Y, Ying C, Lateef SA, Ihara I, Iwasaki M, Inoue R, et al. Anaerobic co-digestion of sugar beet tops silage and dairy cow manure under thermophilic condition. J Mat Cycles Waste Manage. 2015;17:540–6.
Önal H, Núñez HM. An economic analysis of transportation fuel policies in Brazil 2014.
Öner C, Altun Ş. Biodiesel production from inedible animal tallow and an experimental investigation of its use as alternative fuel in a direct injection diesel engine. Appl Energ. 2009;86:2114–20.
Ong HC, Mahlia TMI, Masjuki HH, Norhasyima RS. Comparison of palm oil, *Jatropha curcas* and *Calophyllum inophyllum* for biodiesel: a review. Renew Sust Energ Rev. 2011;15:3501–15.
Onuki S, Koziel JA, Jenks WS, Cai L, Rice S, Leeuwen JH. Optimization of extraction parameters for quantification of fermentation volatile by-products in industrial ethanol with solid-phase microextraction and gas chromatography. J Institute Brewing. 2016;122:102–9.
Orsi F, Muratori M, Rocco M, Colombo E, Rizzoni G. A multi-dimensional well-to-wheels analysis of passenger vehicles in different regions: primary energy consumption, CO_2 emissions, and economic cost. Appl Energ. 2016;169:197–209.
Pfenninger S, Keirstead J. Renewables, nuclear, or fossil fuels? Scenarios for Great Britain's power system considering costs, emissions and energy security. Appl Energ. 2015;152:83–93.
Pimentel D, Marklein A, Toth MA, Karpoff MN, Paul GS, McCormack R, et al. Food versus biofuels: environmental and economic costs. Human Ecol. 2009;37:1–12.
Poitrat E. The potential of liquid biofuels in France. Renew Energ. 1999;16:1084–9.
Popovich CA, Damiani C, Constenla D, Martínez AM, Freije H, Giovanardi M, et al. *Neochloris oleoabundans* grown in enriched natural seawater for biodiesel feedstock: evaluation of its growth and biochemical composition. Bioresour Technol. 2012;114:287–93.
Popovicheva O, Engling G, Lin KT, Persiantseva N, Timofeev M, Kireeva E, et al. Diesel/biofuel exhaust particles from modern internal combustion engines: microstructure, composition, and hygroscopicity. Fuel. 2015;157:232–9.
Popp J, Lakner Z, Harangi-Rákos M, Fári M. The effect of bioenergy expansion: food, energy, and environment. Renew Sust Energ Rev. 2014;32:559–78.

Pöschl M, Ward S, Owende P. Evaluation of energy efficiency of various biogas production and utilization pathways. Appl Energ. 2010;87:3305–21.

Prasad S, Singh A, Joshi H. Ethanol as an alternative fuel from agricultural, industrial and urban residues. Resour Conserv Recycl. 2007;50:1–39.

Ragit SS, Mohapatra SK, Kundu K, Gill P. Optimization of neem methyl ester from transesterification process and fuel characterization as a diesel substitute. Biomass Bioenergy. 2011;35:1138–44.

Rahman MM, Hasan MM, Paatero JV, Lahdelma R. Hybrid application of biogas and solar resources to fulfill household energy needs: A potentially viable option in rural areas of developing countries. Renew Energ. 2014;68:35–45.

Rakopoulos DC, Giakoumis EG, Papagiannakis RG, Kyritsis DC. Influence of properties of various common bio-fuels on the combustion and emission characteristics of high-speed DI (direct injection) diesel engine: vegetable oil, bio-diesel, ethanol, n-butanol, diethyl ether. Energy. 2014;73:354–66.

Ramadhas AS, Jayaraj S, Muraleedharan C. Use of vegetable oils as I.C. engine fuels-A review. Renew Energ. 2004;29:727–42.

Ramadhas AS, Jayaraj S, Muraleedharam C. Biodiesel production from high FFA rubber seed oil. Fuel. 2005;84:335–40.

Rao PV, Baral SS, Dey R, Mutnuri S. Biogas generation potential by anaerobic digestion for sustainable energy development in India. Renew Sust Energ Rev. 2010;14:20–2094.

Rashedul HK, Masjuki HH, Kalam MA, Teoh YH, How HG, Fattah IR. Effect of antioxidant on the oxidation stability and combustion-performance-emission characteristics of a diesel engine fueled with diesel–biodiesel blend. Energ Convers Manage. 2015;106:849–58.

Rasi S, Veijanen A, Rintala J. Trace compounds of biogas from different biogas production plants. Energy. 2007;32:1375–80.

Ratledge C. Fatty acid biosynthesis in microorganisms being used for single cell oil production. Biochimie. 2004;86:8–815.

Ratledge C. Yeast physiology-a micro-synopsis. Bioproc Eng. 1991;6:195–203.

Ray AE, Li C, Thompson VS, Daubaras DL, Nagle NJ, Hartley DS. Biomass blending and densification: impacts on feedstock supply and biochemical conversion performance. Biomass Volume Estimat Valoriz Energ. InTech; 2017. doi:10.5772/67207.

Razaviarani V, Buchanan ID. Anaerobic co-digestion of biodiesel waste glycerin with municipal wastewater sludge: microbial community structure dynamics and reactor performance. Bioresour Technol. 2015;182:8–17.

Ren D, Deng Y, Handoko AD, Chen CS, Malkhandi S, Yeo BS. Selective electrochemical reduction of carbon dioxide to ethylene and ethanol on copper (I) oxide catalysts. ACS Catalysis. 2015;5:2814–21.

Renaud SM, Parry DL, Thinh LV, Kuo C, Padovan A, Sammy N. Effect of light intensity on the proximate biochemical and fatty acid composition of Isochrysis sp. and *Nannochloropsis oculata* for use in tropical aquaculture. J Appl Phycolo. 1991;3:43–53.

Reuter M, Buchwitz M, Hilboll A, Richter A, Schneising O, Hilker M, et al. Decreasing emissions of NO_x relative to CO_2 in East Asia inferred from satellite observations. Nature Geosci. 2014;7:792–5.

Ribéreau-Gayon P, Dubourdieu D, Donéche B, Lonvaud A, Schneising O, Hilker M, et al. Biochemistry of alcoholic fermentation and metabolic pathways of wine yeasts. In: Handbook of enology: the microbiology of wine and vinifications. New York: Wiley; 2006. PP. 53–77.

Riggio V, Comino E, Rosso M. Energy production from anaerobic codigestion processing of cow slurry, olive pomace and apple pulp. Renew Energ. 2015;83:280–93.

Rodriguez C, Alaswad A, Benyounis KY, Olabi AG. Pretreatment techniques used in biogas production from grass. Renew Sust Energ Rev. 2017;68:1193–204.

Rothman H, Greenshields R, Rosillo CF. Energy from alcohol: the Brazilian experience. University Press of Kentucky; 2015.

Roukas T. Ethanol production from carob pod extract by immobilized *Saccharomyces cerevisiae* cells on the mineral kissiris. Food Biotechnol. 1995;9:175–88.

Sagagi B, Garba B, Usman N. Studies on biogas production from fruits and vegetable waste. Bayero J Pure Appl Sci. 2009;2:115–8.

Saka S. Production of biodiesel: current and future technology. In: JSPS/VCO Core University Program Seminar, Universiti Sains Malaysia, September 2005.

Saravanan S, Nagarajan G, Rao GLN. Investigation on nonedible vegetable oil as a compression ignition engine fuel in sustaining the energy and environment. J Renew Sust Energ. 2010;2: 1–8.

Sarin R, Sharma M, Khan AA. Studies on *Guizotiaabyssinica L.* oil: Biodiesel synthesis and process optimization. Bioresour Technol. 2009;100:4187–92.

Sarma A, Konwer D, Bordoloi PK. A comprehensive analysis of fuel properties of biodiesel from koroch seed oil. Energ Fuels. 2005;19:656–7.

Sarris D, Giannakis M, Philippoussis A, Komaitis M, Koutinas AA, Papanikolaou S. Conversions of olive mill wastewater-based media by *Saccharomyces cerevisiae* through sterile and non-sterile bioprocesses. J Chem Technol Biotechnol. 2013;88:958–69.

Sarris D, Matsakas L, Aggelis G, Koutinas AA, Papanikolaou S. Aerated vs non-aerated conversions of molasses and olive mill wastewaters blends into bioethanol by *Saccharomyces cerevisiae* under non-aseptic conditions. Ind Crops Prod. 2014;56:83–93.

Sarris D, Papanikolaou S, Biotechnological production of ethanol: Biochemistry, processes and technologies. Eng Life Sci. 2016.

Satyanarayana M, Muraleedharan C. Biodiesel production from vegetable oils: a comparative optimization study. J Biobas Material Bioenerg. 2009;3:335–41.

Satyanarayana M, Muraleedharan C. A comparative study of vegetable oil methyl esters (biodiesels). Energy. 2011;36:2129–37.

Scano EA, Asquer C, Pistis A, Ortu L, Demontis V, Cocco D. Biogas from anaerobic digestion of fruit and vegetable wastes: experimental results on pilot-scale and preliminary performance evaluation of a full-scale power plant. Energ Convers Manage. 2014;77:22–30.

Searchinger T, Heimlich R, Houghton RA, Dong F, Elobeid A, Fabiosa J, et al. Use of US croplands for biofuels increases greenhouse gases through emissions from land-use change. Science. 2008;319:1238–40.

Shahid EM, Jamal J. Production of biodiesel: a technical review. Renew Sust Energ Rev. 2011;15:4732–45.

Shapouri H, Duffield JA, Wang M. The energy balance of corn ethanol revisited. Trans ASAE. 2003;46:959–68.

Sharma YC, Singh B. Development of biodiesel: current scenario. Renew Sust Energ Rev. 2009;13:1646–51.

Shirazi MJA, Bazgir S, Shirazi MMA. Edible oil mill effluent; a low-cost source for economizing biodiesel production: electrospun nanofibrous coalescing filtration approach. Biofuel Res J. 2014;1:39–42.

Shukla SK, Thanikal JV, Haouech L, Patil, SG, Kumar V. Critical evaluation of algal biofuel production processes using waste water in algal biofuels. Springer International Publishing. 2017;189–225.

Sims RE, Mabee W, Saddler JN, Taylor M. An overview of second generation biofuel technologies. Bioresour Technol. 2010;101:1570–80.

Singh SP, Singh D. Biodiesel production through the use of different sources and characterization of oils and their esters as the substitute of diesel: a review. Renew Sust Energ Rev. 2010;14:200–16.

Smith K. Biofuels, air pollution, and health: a global review. Springer Science & Business Media. 2013.

Soroushian F, Shang Y, Whitman EJ, Garza G, Zhang Z. Development and application of biological H2S scrubbers for treatment of digester gas. P Water Enviro Fed. 2006;9:3541–7.

Sridevi VD, Rema T, Srinivasan SV. Studies on biogas production from vegetable market wastes in a two-phase anaerobic reactor. Clean Technol Environ Policy. 2015;17:1689–97.

Stanbury PF, Whitaker A, Hall SJ. Principles of fermentation technology, Elsevier. 2013.

Tabatabaei M, Karimi K, Horváth IS, Kumar R. Recent trends in biodiesel production. Biofuel Res J. 2015;7:258–67.

Tan HW, Aziz AA, Aroua MK. Glycerol production and its applications as a raw material: a review. Renew Sust Energ Rev. 2013;27:118–27.

Teo SH, Islam A, Yusaf T, Taufiq-Yap YH. Transesterification of Nanno chloropsis oculata microalga's oil to biodiesel using calcium methoxide catalyst. Energy. 2014;78:63–71.

Thiyagarajan S, Edwin Geo V, Martin LJ, Nagalingam B. Carbon dioxide (CO_2) capture and sequestration using biofuels and an exhaust catalytic carbon capture system in a single-cylinder CI engine: an experimental study. Biofuels. 2017;1–10.

Tilman D, Socolow R, Foley JA, Hill J, Larson E, Lynd L, et al. Beneficial biofuels-the food, energy, and environment trilemma. Science. 2009;325:270–1.

Timilsina GR. Biofuels in the long run global energy supply mix for transportation. Philos Trans Royal Soc London A Math Phys Eng Sci. 2014;372:20120323.

Tsoutsos TD, Tournaki S, Paraíba O, Kaminaris SD. The used cooking oil-to-biodiesel chain in Europe assessment of best practices and environmental performance. Renew Sust Energ Rev. 2016;54:74–83.

Uddin M, Rashid MM, Nithe NA, Rony JI. Performance and cost analysis of diesel engine with different mixing ratio of raw vegetable oil and diesel fuel. 2016.

Van Uytvanck PP, Hallmark B, Haire G, Marshall PJ. Impact of biomass on industry: using ethylene derived from bioethanol within the polyester value chain. ACS Sust Chem Eng. 2014;2:1098–105.

van Eijck J, Romijn H, Smeets E, Bailis R, Rooijakkers M, Hooijkaas N, et al. Comparative analysis of key socio-economic and environmental impacts of smallholder and plantation based jatropha biofuel production systems in Tanzania. Biomass Bioenerg. 2014a;61:25–45.

van Eijck, J, Batidzirai B, Faaij A. Current and future economic performance of first and second generation biofuels in developing countries. Appl Energ. 2014b;135:115–41.

Vázquez Ojeda M, Segovia Hernández JG, Hernández S, Hernández Agurri, Kiss AA. Design and optimization of an ethanol dehydration process using stochastic methods. Sep Purific Technol. 2013;105:90–7.

Walker G, Simcock N, Day R. Necessary energy uses and a minimum standard of living in the United Kingdom: energy justice or escalating expectations? Energ Res Soc Sci. 2016;18: 129–38.

Ware A, Power N. Biogas from cattle slaughterhouse waste: Energy recovery towards an energy self-sufficient industry in Ireland. Renew Energ. 2016;97:541–9.

Wellinger A, Murphy JD, Baxter D. The biogas handbook: science, production and applications. Elsevier; 2013.

Wesseler J, Drabik D. Prices matter: analysis of food and energy competition relative to land resources in the European Union. NJAS-Wageningen J Life Sci. 2016;77:19–24.

Westbrook CK, Naik CV, Herbinet O, Pitz WJ, Mehl M, Sarathy SM, et al. Detailed chemical kinetic reaction mechanisms for soy and rapeseed biodiesel fuels. Combust Flame. 2011;158:742–55.

Weusthuis RA, Pronk JT, van den Broek P, van Dijken J. Chemostat cultivation as a tool for studies on sugar transport in yeasts. Microbiol Rev. 1994;58:616–30.

Wolters N, Schabronath C, Schembecker G, Merz J. Efficient conversion of pretreated brewer's spent grain and wheat bran by submerged cultivation of *Hericium erinaceus*. Bioresour Technol. 2016;222:123–9.

Wyatt VT, Hess MA, Dunn RO, Foglia TA, Haas MJ, Marmer WN. Fuel properties and nitrogen oxide emission levels of biodiesel produced from animal fats. J Am Oil Chem Soc. 2005;82:58–91.

Yanai T, Dev S, Han X, Zheng M, Tjong J. Impact of fuelling techniques on neat n-butanol combustion and emissions in a compression ignition engine. SAE Intl J Engines. 2015;8: 735–46.

Yusuf NNAN, Kamarudin SK, Yaakub Z. Overview on the current trends in biodiesel production. Energ Convers Manage. 2011;52:2741–51.

Zabeti M, Daud WMAW, Aroua MK. Activity of solid catalysts for biodiesel production: A review. Fuel Process Technol. 2009;90:770–7.

Zaborsky OR. Chemicals from renewable resources: an endorsement for biotechnology. Enzyme Microbiol Technol. 1982;4:364–5.

Zamora F. Biochemistry of alcoholic fermentation, wine chemistry and biochemistry. New York: Springer; 2009. PP. 3–26.

Zarkadas IS, Sofikiti AS, Voudrias EA, Pilidis GA. Thermophilic anaerobic digestion of pasteurised food wastes and dairy cattle manure in batch and large volume laboratory digesters: focussing on mixing ratios. Renew Energ. 2015;80:432–40.

Zhang H, Zhou Q, Chang F, Pan H, Liu XF, Li H, et al. Production and fuel properties of biodiesel from *Firmiana platanifolia* L.f. as a potential non-food oil source. Indus Crop Prod. 2015;76:768–71.

Zhu J, Zheng Y, Xu F, Li Y. Solid-state anaerobic co-digestion of hay and soybean processing waste for biogas production. Bioresour Technol. 2014;154:240–7.

Zieminski K, Kowalska-Wentel M. Effect of enzymatic pretreatment on anaerobic co-digestion of sugar beet pulp silage and vinasse. Bioresour Technol. 2015;180:274–80.

Chapter 2
Climatic Changes Impact on Water Availability

Ijaz Bano and Muhammad Arshad

Abstract Climate changes refer to long term changes in the global climate that is interconnected system of universe. This system includes sun, earth, oceans, wind, rain, snow, forests, deserts, savannas and all human activities. Availability of potable and non potable water major relies on rainfall and oceans reserves. Globally, rainfall pattern has changed because of rise in temperature as a result of climatic changes. As a result of climatic changes, water demand and availability of freshwater resources are affecting at local, regional and global levels. In addition, increased population growth, resulting urbanization, consequent industrialization, competing demands for water and altered socio-economic conditions have put additional pressure on the water supply. While, the demand increases quickly during prolonged hot spells. However an impact that we have on the environment with CO_2 emissions reaches their highest in the industrial time. Consequently, the prediction of potential impacts of climatic changes on water availability and demand is crucial to take suitable steps to mitigate the adverse impacts of climate change on water availability. This chapter will appraise the impact of climate change on water availability particularly in dry regions of the world. This study will guide water management systems to carry out effective and efficient long-term planning for the sustainable supply of water under different weather conditions and population scenarios.

Keywords Climate change · Global warming · Water pollution · Fresh water

2.1 Introduction

"NATURE! We are surrounded and embraced by her: powerless to separate ourselves from her, and powerless to penetrate beyond her". Consequently water is the lifeblood of our Nature (Rahaman 2012). Water cannot be separated from our-

I. Bano (✉) · M. Arshad
Jhang-Campus, University of Veterinary and Animal Sciences, Lahore, Pakistan
e-mail: ijaz.bano@uvas.edu.pk

M. Arshad
e-mail: Muhammad.arshad@uvas.edu.pk

M. Arshad (ed.), *Perspectives on Water Usage for Biofuels Production*,
https://doi.org/10.1007/978-3-319-66408-8_2

selves, it is vital for our very continued existence. Much interesting that water needs not to follow any political borders. How we use and manage our valuable and limited water capitals is very critical. Around the globe in many regions, human being has shown mismanagement and polluted the limited water reservoirs.

Drinking water is the major and current issue of this world (Richardson and Ternes 2011). Availability of water; safe to human health is alarming (Schwarzenbach et al. 2010). Next world war seems to be over water among superpowers on earth. Fresh water reservoirs are depleting day by day (Berger et al. 2014). Our earth came to exist many billion years ago (Swimme and Berry 1994). We cannot exactly predict the earth's daily weather was like in any particular place on any particular day billions of years ago (Palmer 2012). We only know lot about what the earth's climate was because of clues that remains in rocks, ice, trees and fossils. According to these clues, the earth's climate has changed many times. Initially, this planet was covered with ice, though some warmer periods have been detected too. Over the past decades, temperatures and carbon dioxide levels in the atmosphere have increased due to fossil fuel utility (Flannery 2006.). We however know that industrialization cannot occur without changing climate. But, global warming due to industrialization is resulted in extreme climatic events such as droughts, floods and heat waves (Beniston et al. 2007; Kovats et al. 2014; Bucak et al. 2017). Mediterranean regions suffering water scarcity and droughts due to extensive use of water in industries (Chenini 2010). Also, significant changes in freshwater availability are expected to occur with the ongoing climate changes.

Climatic projections for the Mediterranean region have predicted a significant decrease in precipitation and enhanced temperatures. This is resulting in increased dry days and more frequent heat waves (Erol and Randhir 2012; Christensen et al. 2013; Giannakopoulos et al. 2009). Consequently, water evaporation from sea and land surfaces due to higher temperatures may lead to a further decrease in water availability. These situations add problems to existing water scarcity all over the region (Calbó 2010).

Water planners and managers around the globe have main focus on climatic changes. It is important to change water managements by altering the availability of freshwater resources and water demand patterns (Jeuland and Whittington 2014). Due to industrialization and fossil fuel burning, more of the rise in temperature and changes in rainfall pattern are expected to occur in many part of the earth (Chadwick et al. 2014). Such changes are negatively affecting freshwater balance at local, regional and global levels. These changes will make safe water supply a big challenge to water managers (Chang 2007). Some other factors in addition of global warming are likely to affect fresh water availability in future which includes increasing population growth, increased water demand (McDonald et al. 2011), water pollution (Peters and Meybeck 2000) and rapid urbanization. Population in different continents, irrigatable land, and sustainable water supplies has presented in Table 2.1.

Alternative and suitable managements are necessary to keep balance between water supply and demand in current changing climate (Rosenzweig et al. 2004). Water demand also varies with climatic variables and seasons. During hot days and

Table 2.1 Population in different continents, irrigatable land, and sustainable water supplies (Vorosmarty et al. 2000)

Area	Population (millions)		Irrigated cropland	Observed water supply	Water supply required in 2025
	1985	2025	1000 km^2	km^3 year^{-1}	km^3 year^{-1}
Africa	543	1440	118	4520	4100
Asia	2930	4800	1690	13,700	13,300
Australia	22	33	26	714	692
Europe	667	682	273	2770	2790
North America	395	601	317	5890	5870
South America	267	454	95	11,700	10,400
Worldwide	4830	8010	2520	39,300	37,100

low rainfall, water demands increases for normal human activities and vice versa (Shahid 2011). Influence of the climatic variables on water availability has been reported by many studies. For example, rainfall is the one of key marker for water demand variables in Kathmandu, Nepal (Babel et al. 2007) for the prediction of domestic water. Gato et al. (2007) found that temperature and rainfall have statistically a significant correlation with water usage in Melbourne, Australia. In a review of the significant variables of domestic water demand. Babel and Shinde (2011) concluded that future water demand in Bangkok could be significantly affected by climate change, as meteorological variables such as temperature, rainfall and relative humidity have a considerable influence on longer term demand projection. Xiao-jun et al. (2015) found that future water demand in Yulin City, Northwest China would rise due to changes in climatic conditions, especially rise in temperature.

In Australian cities, water supply is more vulnerable to changes in climatic conditions as it is highly dependent on rainfall and storage capacity of surface water reservoirs. However, rainfall in Australia is highly variable (Sahin et al. 2013) and about 50–70% of the country is in the semiarid and arid regions where rainfall is very low (Zaman et al. 2012). During the droughts in Australia (2003–2009), most of the major reservoirs reached critical low water levels, thereby putting water supply at risk. The annual average temperature in Australia increased by 0.9 °C from 1910 to 2011 which is higher than the global average increase of 0.7 °C for the same period (Cleugh et al. 2011). Most of this increment in temperature has occurred since the 1950s, with the highest increment in the eastern part of Australia by 2 °C and the lowest increment in the northwest part by −0.4 °C.

Like many things in life, a little global warming is critical for your survival, while too much will destroy you. Global warming is the increasing rise in the earth's temperature through the increase in greenhouse gases (Ehrlich and Ehrlich 1991). The earth's temperature is regulated largely by gases that trap heat in the earth's atmosphere. Often this is referred to as the greenhouse effect. This warming

Table 2.2 Relative greenhouse impact by different greenhouse gases

Greenhouse gases	Relative impact	Reference
CO_2	1	Rodhe (1990)
CH_4	21	
N_2O	310	
SF_4	23,900	
PFCs	6500–9200	
HFCs	140–11,700	

has been critical in providing the earth with the right temperature that we need to survive. This trapping of the heat allows the earth's temperature to be in the range to support life (Lacis et al. 2010). Relative greenhouse impact by different greenhouse gases is presented in Table 2.2. Unfortunately, due to man's recent activities, we have increased the concentration of specific greenhouse gases and are now responsible for ever increasing temperatures. The greenhouse gases include such gases as carbon dioxide, methane (Mitsch et al. 2013), halocarbons (Lim et al. 2017), ozone (Fann et al. 2015) and nitrous oxide (Hartmann et al. 2013). They have the ability to absorb the heat radiated from the earth's surface.

It has been difficult to confirm that our effect on global warming is becoming critical because there always has been a natural variability in earth's temperature. However, it is now clear that human activities have been causing an increase in the earth's temperature. It is estimated that most of the current global warming has occurred since the mid-20th century (Knutson et al. 2016). Carbon dioxide has risen about 30% since the late 1800s (Rhein et al. 2013). This is a result of the burning of coal, oil and natural gas and the destruction of forests around the world.

2.2 Climatic Changes Over Bio-Planet

Climate change refers to a general change in climate pattern which, include temperature, rainfall, winds and other related factors, while global warming and cooling refers to change in the global average surface temperature. Increase in average global temperature is the major factor of climatic changes (IPCC 2014). Predominantly, human activities are major contributors of global warming because of increased industrialization and greenhouse gases.

2.2.1 Role of Greenhouse Gases

Greenhouse gases include carbon dioxide (CO_2), methane (CH_4), nitrous oxide (N_2O), fluorinated industrial gases such as hydrofluorocarbons (HFCs), perfluorocarbons (PFCs) and sulphur hexafluoride (SF_6). Among them, methane is 20 times as potent a greenhouse gas as carbon dioxide. Water vapors also included in greenhouse gases (Stocker 2014). Natural and industrial greenhouse impact on biosphere has been shown in Fig. 2.1. A planet with increased temperature steer the climate which in turn affect the weather in various ways. So, the term "greenhouse" is used in conjunction with the phenomenon known as the greenhouse effect. The greenhouse effect is the causing of global warming on earth as certain gases in the atmosphere trap sun energy. Sun-rays are the key drivers of earth's weather and climate as it heats the surface of planet. In turn, the earth bounces back the energy radiations into the space. Some greenhouse gases trap some of the outgoing energy, retaining heat somewhat like the glass panels of a greenhouse. These gases are therefore known as greenhouse gases

Many of these greenhouse gases are actually important for life on earth, as without greenhouse gases, heat would not stay and as a result, the average temperature of the earth would be colder. In contrary, with increasing greenhouse effect, more heat gets trapped and the earth might become less suitable for humans, plants and animals. Carbon dioxide, though not the most potent of greenhouse gases, is the most significant one. Natural cycle of the greenhouse effect has affected by human activities. Climatic changes due to natural and human activities have been presented in Fig. 2.2. The natural fluctuation of carbon through the earth system, anthropogenic activities, particularly fossil fuel burning and deforestation are also releasing carbon dioxide into the atmosphere. In addition, fossil fuels like coal and oil use for energy purpose in transportation, heating, cooking, electricity, and manufacturing are effectively moving carbon rapidly into the atmosphere than is being removed. Dry wood contains 50% carbon which humans are converting into carbon dioxide by deforestation. The result is that humans are adding increasing amounts of extra carbon dioxide into the atmosphere.

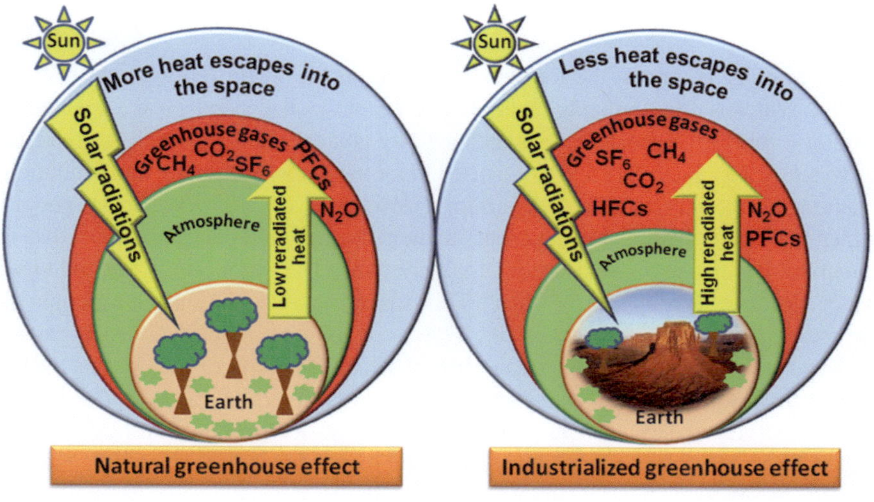

Fig. 2.1 Natural and industrial greenhouse impact on biosphere

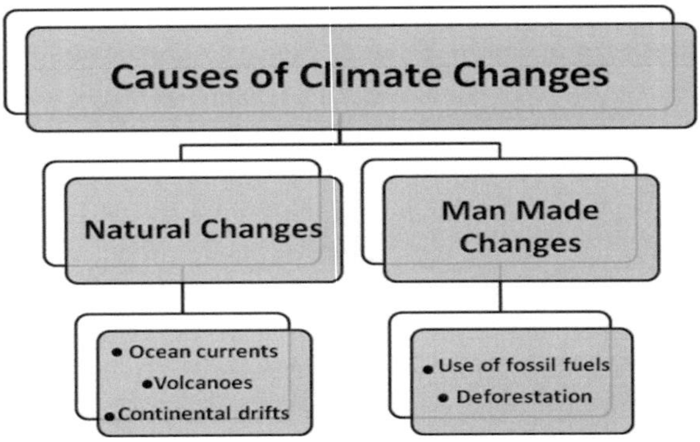

Fig. 2.2 Climatic changes due to natural and human activities

Climate has changed throughout the history of earth. However, past warming spells does mean today's warming is also natural. Recent warming has been shown to be due to human industrialization processes. Satellite observations show that the arctic sea ice is decreasing, and projections for the rest of the century predict even more shrinkage. Climate change impacts the rising of sea levels. Climate scientists have predicted about 20 cm rise in average sea level during past 100 years and expect it to rise more and more rapidly in the next 100 years as a part of climate change impacts. New York and such other coastal cities are already facing high flooding events and by 2050 many such cities expecting more and may require seawalls to survive. Estimates vary, but conservatively sea levels are expected to rise 1–4 ft (30 to 100 cm), enough to flood many small Pacific island states, famous beach resorts and coastal cities.

2.3 Climate Changes and Water Availability Issues

Water scarcity is among the top most concerns of the world (Brown and Halweil 1998). Water is life as plants, animals and humans all depend on the invaluable natural resources. Almost every human activity you can think of involves some use of water (Kingsolver 2010). Water is used in moving waste, cleaning and sanitation, manufacturing, construction and farming (Howard et al. 2003). For many people, water has never been a big problem in their lives. This is because they live in communities that have good water supply systems. They turn on the tap and clean water flows, everyday of the year. This makes it very difficult for people to appreciate how precious water is for. Water covers more than 70% of the earth's surface but less than 3% of it is fresh water (Gleick 2014). Major bulk of fresh water is entrapped to snowfields and glaciers which are not in easy access. The rest of water reserves as seas, oceans cannot be used as fresh water. A tiny fraction (0.014%) is as surface water in the form of rivers, lakes etc. Naturally, the 3% should be enough for all humans and animals on earth, but unfortunately,

Table 2.3 Possible impacts of climate change on water resources (Zimmerman et al. 2008)

Phenomena	Chances or future tendencies	Major effects on water resources
Plain land areas possibly will have long warmer and fewer cold days	Nigh on	Water resources relying on snowmelt will be effected and disrupt the water supplies
Frequency of warmer spells in summer season and heat waves will be increased in most of the land areas	Very well expected	Water requirements will be increased
Occurrence of heavy rainfall happenings will be increased	Almost immediately possible	Deterioration of water quality at surface and groundwater
Many areas possibly will be affected by drought	To be expected	Stress on water resources will very high
Tendency of tropical cyclones may possibly be increased	Probability is high	Power outbreaks and interruption of water supply
Sea levels can be extremely high	Very nearly	Freshwater will reduce due to saltwater invasion

many factors have caused a major upset in the flow and use of fresh water and has caused massive crisis in many regions of the earth.

If climate in a given region were to become warmer and drier, water availability would decrease and water demand would increase, especially demand for irrigation and electric power production, the largest users of water (Schewe et al. 2014). Probable impacts of climate change on water resources are presented in Table 2.3. Lower river flows resulting from drier conditions could affect adversely in stream uses such as hydropower, navigation, aquatic ecosystems, wildlife habitat and recreation. Lower stream flow and lower lake levels could cause new power plants to locate in coastal areas in order to obtain a water source that is reliable and that may be used without violation of thermal restrictions (EPA 1989).

2.3.1 *Water Pollution*

The air is polluted by burning fossil fuels in industries, power plants, transports etc. (Lee et al. 2014). When sulphur and nitrogen oxides escape to atmosphere these pollute the air and produce toxic clouds. These toxic clouds in turn pollute the land surface in the form of acid rain which ultimately pollutes the river water which now cannot be used for drinking purpose (Lamarque et al. 2013). Industries cause huge water pollution with their activities. These come mainly from sulphur that is a non-metallic substance and is harmful to marine life. Asbestos pollutant has cancer-causing properties. When inhaled, it can cause illnesses such as asbestosis and some types of cancer. Lead and Mercury are metallic elements and can cause environmental and health problems for humans and animals. It is also poisonous. It is usually very hard to clean it up from the environment once it gets into it because it is non-biodegradable (de Vries et al. 2013). Nitrates and Phosphates are found in fertilizers, and are often washed from the soils to nearby water bodies. They can cause eutrophication, which can be very problematic to marine environments. Oils form a thick layer on the water surface because they do not dissolve in water. This can stop marine plants receiving enough light for photosynthesis. It is also harmful to fish and marine birds. A classic example is the BP (British petroleum) oil spill in 2012 which killed thousands of animal species. Routine shipping, run-offs and dumping of oils on the ocean surfaces happen every day. Oil spills make up about 12% of the oil that enters the ocean. Oil spills cause major problems, and can be extremely harmful to local marine wildlife such as fish, birds and sea otters and other aquatic life. Because oil does not dissolve, it stays on the water surface and suffocates fish. Oil also gets caught in the feathers of seabirds, making it difficult for them to fly. Some animals die as a result. A change in the chemical, physical, biological, and radiological quality of water that is injurious to its uses. Thus, the discharge of toxic chemicals from industries or the release of human or livestock waste into a nearby water body is considered pollution.

The contamination of ground water of water bodies like rivers, lakes, wetlands, estuaries, and oceans can threaten the health of humans and aquatic life. Sources of water pollution may be divided into two categories. (i) Point-source pollution, in which contaminants are discharged from a discrete location. Sewage outfalls and oil spills are examples of point-source pollution. (ii) Non-point-source or diffuse pollution, referring to all of the other discharges that deliver contaminants to water bodies. Acid rain and unconfined runoff from agricultural or urban areas falls under this category (Black et al. 2014). The principal contaminants of water include toxic chemicals, nutrients, biodegradable organics, and bacterial and viral pathogens. Water pollution can affect human health when pollutants enter the body either via skin exposure or through the direct consumption of contaminated drinking water and contaminated food. Prime pollutants, including DDT and polychlorinated biphenyls (PCBs), persist in the natural environment and bioaccumulation occurs in the tissues of aquatic organisms (Mckinney et al. 2015). These prolonged and persistent organic pollutants are transferred up the food chain and they can reach levels of concern in fish species that are eaten by humans. Moreover, bacterial and viral pathogens can pose a public health risk for those who drink contaminated water or eat raw shellfish from polluted water bodies.

Contaminants have a significant impact on aquatic ecosystems. Enrichment of water bodies with nutrients (principally nitrogen and phosphorus) can result in the growth of algae and other aquatic plants that shade or clog streams. If wastewater containing biodegradable organic matter is discharged into a stream with inadequate dissolved oxygen, the water downstream of the point of discharge will become anaerobic and will be turbid and dark. Settle able solids will be deposited on the streambed, and anaerobic decomposition will occur. Over the reach of stream where the dissolved-oxygen concentration is zero, a zone of putrefaction will occur with the production of hydrogen sulfide (H_2S), ammonia (NH_3), and other odorous gases. Because many fish species require a minimum of 4–5 mg of dissolved oxygen per liter of water, they will be unable to survive in this portion of the stream. Direct exposures to toxic chemicals are also a health concern for individual aquatic plants and animals. Chemicals such as pesticides are frequently transported to lakes and rivers via runoff, and they can have harmful effects on aquatic life. Toxic chemicals have been shown to reduce the growth, survival, reproductive output, and disease resistance of exposed organisms. These effects can have important consequences for the viability of aquatic populations and communities.

2.3.2 Wastewater Discharges

Wastewater discharges are most commonly controlled through effluent standards and discharge permits. Under this system, discharge permits are issued with limits on the quantity and quality of effluents. Water-quality standards are sets of qualitative and quantitative criteria designed to maintain or enhance the quality of receiving waters. Criteria can be developed and implemented to protect aquatic life against acute and chronic effects and to safeguard humans against deleterious health effects, including cancer.

2.3.3 Water and Flood Issues in Pakistan

The looming threat of water scarcity is an issue that is rarely talked about in Pakistani politics, and yet it constitutes one of the biggest challenges to Pakistan's survival. With a projected population of 263 million in the year 2050 (United Nations 2012), Pakistan needs to put serious thought into how it will provide adequate water for agriculture, industry, and human consumption in the face of rapidly dwindling reserves. The Himalayan glacier, whose ice melt replenishes the Indus River's annual freshwater, is receding by about one meter the approximate equivalent of 3.3 ft per year due to global warming (Simi 2009). This phenomena has had a staggering impact on Pakistan's water availability. In just 1950, Pakistan had around 5,000 m^3 per capita per year of freshwater resources. In 2002, its supplies shrunk to only 1,500 m^3. To put that number in perspective, around 1,000 m^3 is when a country is declared water scarce (Michael 2009).

Unfortunately, Pakistan's water woes do not end with just scarcity. The War on Terror, as well as the 2010 flood, has displaced two million people from Pakistan's countryside, many of whom have flocked to urban centers such as Karachi. As a result, Pakistan must also increase water availability and sanitation in urban centers to accommodate this massive influx of people, in addition to tackling its water scarcity issue. One figure states that around 40–55 million Pakistanis do not have regular access to drinking water and around 630 Pakistani children die each day to the waterborne illness of diarrhea (Michael 2009).

The mismanagement of water will have its biggest impact on Pakistan's agricultural sector. According to the World Bank, 43% of Pakistan's employment is in the agricultural sector (WDI 2013). This prosperous industry relies on the single largest contiguous irrigation system in the world. While this is an impressive feat, Pakistan also fosters one of the lowest crop yields per unit of water in the world. This is alarming because Pakistan uses a whopping 97% of its water resources on its agriculture industry (Simi 2009).

As the previous examples demonstrate, Pakistan's water issues are multi-dimensional. There is no single, all-encompassing problem, but instead multiple, interrelated problems. Therefore, Pakistan needs to completely rethink its entire approach to its water resources. It will take time to implement solutions to these problems, and yet time is in very short supply. It is projected that by about 2035, Pakistan will become water scarce (Simi 2009).

The current status quo is characterized by waste, provincial disputes, corruption, and poor infrastructure. Feisal Khan, Assistant Professor of Economics at Hobart and William Smith Colleges, points out that Pakistan's crop irrigation system is in desperate need of repair. The annual budget for maintenance and repair usually falls well below what is needed. In 2005, the projected budget for repair of Punjab's irrigation infrastructure was estimated at 0.6 billion USD; unfortunately, only 0.2 billion USD was allocated towards maintenance. As such, Pakistan practices a cycle of "Build/Neglect/Rebuild" when it comes to its water infrastructure (Feisal 2009). Currently, around one-third of the water from the irrigation system is lost in delivery due to seepage and malfunctioning watercourses. Still, this number does not seem to faze officials in Islamabad, as Pakistan continues to push for the construction of expensive new dams, such as Diamer Bhasha, instead of renovating its decaying irrigation system (Simi 2009). In fact, there is evidence that dam projects have contributed greatly to waterlogging and soil salinity, making large chunks of agricultural land useless (Kaiser 2009).

Another source of waste comes from farms themselves. Around a quarter of the water delivered from irrigation is wasted from poor farming practices (Simi 2009). Khan blames this on the warabandi system of water management in rural areas. According to this system, each farmer has a specific day to irrigate his or her field. The quantity of water used is irrelevant each farmer pays a flat fee. Although this system was intended to be equitable in the face of water shortages, in reality, farmers who have first access can take a lion's share of the water. Since water is not priced based on usage, there is nothing to discourage waste and overuse. As such, large and powerful farmers have greater access to water from the Indus, which forces small farmers to rely on tube wells to extract groundwater (Feisal 2009). In turn, over-extraction of groundwater negatively affects the salt content of soil, leading to further environmental destruction. This inequality in water distribution also negatively affects crop yields, since small farmers do not have access to adequate water supplies.

Urban areas lack adequate water treatment facilities. As of 2007, only about 7.7% of urban wastewater underwent treatment. Most of the time, household and industrial waste is simply dumped into nearby waterways. This practice has dramatically raised the level of pollution in both groundwater and river systems, and constitutes a major concern for public health.

Lastly, water scarcity has several political implications. Numerous disputes have erupted between Punjab and Sindh over water use and allocation. Although there are provisions within the government which outline the distribution of water between the two provinces, these guidelines are rarely ever met. The current system is particularly hurtful to Sindh, as water from the Indus and its tributaries must pass through Punjab first. Any interference with the Indus River in Punjab adversely affects Sindh. As more and more water is redirected to support Punjab's agricultural industry, there is less water in Sindh for use in consumption, sanitation, and environmental conservation (Feisal 2009). By some accounts, the Sindhi portion of the Indus has shrunk to the size of a mere canal (Michael 2009).

Water pollution is the death of the earth. We should care because a lot of the factors that cause water scarcity are broadening and becoming more complex and uncontrollable. This means if we do nothing in terms of preserving and using it wisely, it is only a matter of time that all regions shall begin to experience water crisis and all the repercussions that come with it. Ozone is a natural gas and is naturally replenished over time. This means if we can do something to balance the natural production with its depletion, there should not be a problem.

Unfortunately, it does not quiet work like that. People ask if we cannot produce our own ozone gas to replenish what is lost in the stratosphere. That's a good question. The sun naturally produces ozone with immense energy and over time. To do the same, we will be looking at using immense energy too, about twice the energy used in the USA. That is just not practical. The only way to do that is to remove the excess chlorine and bromine from the stratosphere. And the only way to do that is to stop making CFCs and several other chemicals. This is why in the 1990s a meeting of the worlds big nations met and agreed to reduce the usage of CFCs and also encouraged other nations to do the same. That was decided in the Montreal Protocol. This is not enough, but at least it was a good starting point. It is always best to talk and discuss problems than to do nothing at all. This is why learning about Ozone depletion, like you are doing, is the most important step towards a safe environment in future.

Finally, Pakistan needs to change its mentality towards conservation. For most of Pakistan's history, the response to water shortages has always been to build dams, redirect rivers, and irrigate the soil. While this may have worked in the past, this engineering-based approach largely ignores the reality that Pakistan is sitting on dwindling reserves. The approach for the coming century must focus on re-education and conservation. Instead of thinking about how and where to build new dams, Pakistan should be thinking about how it can reduce waste in the existing system. Instead of figuring out ways to extract more water from the ground, Pakistan should be figuring out how to recycle its water and make the most of each drop. In a country that has the largest contiguous irrigation system in the world, the main obstacle is not that the system is not large enough, but that it has not been streamlined for efficiency.

2.4 Human Effects on Water Quality and Quantity

Due to human activity, there is a direct impact effect on the hydrologic cycle and the land's physical, chemical, and biological characteristics are changed (Carpenter et al. 2011). Urbanization, transportation, irrigation, deforestation, land drainage, channelization and mining change water ways and also alter water quality parameters by changing the materials with which the water interacts (Peters and Meybeck 2000).

Water quality is not only altered by changes in water ways, but also through the deposition of many substances and wastes to the landscape. Such activities include application of pesticides, herbicides, and fertilizer. These may also leach to the groundwater and surface water from landfills, mine tailings, and irrigated farmland. Table 2.4 keeps valuable information regarding various physical and biological contaminates effecting water quality. The chemical changes are also linked physical processes, but occurs mainly through the addition of wastes gases, liquids, and solids to the earth. Such activities include on land release of industrial effluents or in waterways. Some human-derived substances, including pesticides, microorganic pollutants, nitric acid, and sulfuric acid from fossil fuel combustion, have been traced everywhere. Their occurrence and distribution is due to long-range transport in the atmosphere (Majewski and Capel 1995). Biological modification comprise of forest management, agriculture, and the inclusion of exotic species.

Human direct needs of water also heavily impact hydrologic pathways through the provision of a specified quality water for various activities for human sustainability as farming, Drinking water supplies, electricity generation, cooling towers of the industries. The water quality from urban areas is complex due to the innumerable sources and pathways (Larsen et al. 2009). Artificial drainage channels in urban areas are present in place of natural water ways. Release of untreated waste water directly to earth surface has caused much water pollution, emphasis has been laid on control of leaching the toxic materials (Line et al. 1999).

Table 2.4 Various physical and biological contaminates effecting water quality (Peters and Meybeck 2000)

Sources of contamination	Major issues	Major control factors
Population	Pathogens micro pollutants Eutrophication	Density and various treatment
Water management	Salinization parasites eutrophication	Hydrology and water balance
Land management	Pesticides nutrients physical changes	Agrochemicals, fertilizers cultivation and mining
Atmospheric transport	Micro pollutant radionuclides	Fossil fuel emissions
Global climate changes	Salinization	Fossil fuel emissions and greenhouse gases

The resources used to sustain various standards of humans living are major cause in deterioration of water quality (Moldan et al. 1997). Humans are altering the land to get these resources. Irrigated agriculture alone is responsible for about 75% of the total water withdrawn from surface water and groundwater sources, and more than 90% of this water is consumed and delivered to the atmosphere by evaporation (Cosgrove and Rijsberman 2014). Water quality degradation depends upon climatic characteristics such as amount and timing of rainfall and associated potential evapotranspiration and the various agrochemicals applied to increase yields.

2.5 Conclusion

Climate change mainly creates threats to any of the country's sustainable development. It also effects water resources management and protection. Water scientists have focused their research on establishing sustainable climatic change policies with multi-sectoral coordinating bodies. A large proportion of the world's population is recently suffering from water stress. Rise of global water demand for production of biofuels greatly outweighs greenhouse gases reduction impact of biofuels. Frequent availability of water is considered an important facet of the larger global climate change question. The chapter contributes to an enhanced understanding of present climatic conditions, observed climate trends and climate vulnerability to fresh water availability. Importance of climatic factor in availability of water to raise food and feedstock has been reviewed.

References

Babel MS, Gupta AD, Pradhan P. A multivariate econometric approach for domestic water demand modeling: an application to Kathmandu, Nepal. Water Resour Manage. 2007;21:573–89.

Babel MS, Shinde VR. Identifying prominent explanatory variables for water demand prediction using artificial neural networks: a case study of Bangkok. Water Resour Manage. 2011;25:1653–76.

Beniston M, Stephenson DB, Christensen OB, Ferro CAT, Frei C, Goyette S, et al. Future extreme events in European climate: an exploration of regional climate model projections. Clim Chang. 2007;81:71–95.

Berger M, van der Ent R, Eisner S, Bach V, Finkbeiner M. Water accounting and vulnerability evaluation (WAVE): considering atmospheric evaporation recycling and the risk of freshwater depletion in water footprinting. Environ Sci Technol. 2014;48:4521–8.

Black BA, Lamarque JF, Shields CA, Elkins-Tanton LT, Kiehl JT. Acid rain and ozone depletion from pulsed Siberian Traps magmatism. Geology. 2014;42:67–70.

Brown LR, Halweil B. China's water shortage could shake world food security. World Watch. 1998;11:10–21.

Bucak T, Trolle D, Andersen HE, Thodsen H, Erdoğan S, Levi EE, et al. Future water availability in the largest freshwater Mediterranean lake is at great risk as evidenced from simulations with the SWAT model. Sci Total Environ. 2017.

Calbo J. Possible climate change scenarios with specific reference to Mediterranean regions. In: Sabater S, Barceló D, editors. Water scarcity in the Mediterranean. Springer, Berlin Heidelberg: 2010;1–13.
Carpenter SR, Stanley EH, Vander Zanden MJ. State of the world's freshwater ecosystems: physical, chemical, and biological changes. Ann Rev Environ Resour. 2011;36:75–99.
Chadwick R, Good P, Andrews T, Martin. Surface warming patterns drive tropical rainfall pattern responses to CO_2 forcing on all timescales. Geophy Res Letter. 2014;41:610–15.
Change C. Synthesis report. Contribution of working groups i, ii and iii to the fourth assessment report of the intergovernmental panel on climate change. 2007.
Chenini F. Implications of water quality on irrigation practices under water scarcity. In: Sabater S, Barceló D, editors. Water scarcity in the mediterranean. Berlin, Heidelberg: Springer; 2010. p. 161–72.
Christensen JH, Kanikicharla KK, Marshall G, Turner J. Climate phenomena and their relevance for future regional climate change. 2013;1217–1308.
Cleugh H, Cleugh H, Smith MS, Battaglia M, Graham P. Climate change: science and solutions for Australia. CSIRO; 2011.
Cosgrove WJ, Rijsberman FR. World water vision: making water everybody's business. Routledge; 2014.
deVries W, Groenenberg JE, Lofts S, Tipping E, Posch M. Critical loads of heavy metals for soils. In: Heavy metals in soils. Springer Netherlands; 2013;211–37.
Ehrlich PR, Ehrlich AH. Healing the planet: strategies for resolving the environmental crisis. Our life-support systems–Energy and the environment–Global warming: the beginning of the end–Ozone: a cautionary tale–Pollution: dead trees and poisoned water–Use and abuse of land and water–Unsustainable agriculture–Risks, costs, and benefits–Escaping the human predicament–Influencing policy; 1991.
Erol A, Randhir TO. Climatic change impacts on the ecohydrology of Mediterranean watersheds. Clim Chang. 2012;114:319–41.
Fann N, Nolte CG, Dolwick P, Spero TL, Brown AC, Phillips S, et al. The geographic distribution and economic value of climate change-related ozone health impacts in the United States in 2030. J Air Waste Manage. 2015;65:570–80.
Feisal K. Water, Governance, and corruption in Pakistan. In: Michael K, Robert Hathaway M, editors. Running on empty: Pakistan's water crisis. Washington, D.C: Woodrow Wilson International Center for Scholars, 2009;82–104.
Flannery TF. The weather makers: how man is changing the climate and what it means for life on earth. Grove Press; 2006.
Gato S, Jayasuriya N, Roberts P. Temperature and rainfall thresholds for base use urban water demand modelling. J Hydrol. 2007;337:364–76.
Giannakopoulos C, Sager PL, Bindi M, Moriondo M, Kostopoulou E, Goodess CM. Climatic changes and associated impacts in the Mediterranean resulting from a 2°C global warming. Glob Planet Chang. 2009;68:209–24.
Gleick PH. The world's water volume 8: the biennial report on freshwater resources (Vol. 8). Island Press; 2014.
Hartmann DL, Tank AMK, Rusticucci M, Alexander LV, Brönnimann S, Charabi YAR, et al. Climate change 2013 the physical science basis: working group I contribution to the fifth assessment report of the intergovernmental panel on climate change. Cambridge University Press; 2013.
Howard G, Bartram J. Domestic water quantity, service level and health. World Health Organization; 2003.
IPCC. Climate Change 2014–Impacts, adaptation and vulnerability: regional aspects. Cambridge University Press; 2014.
Jeuland M, Whittington D. Water resources planning under climate change: assessing the robustness of real options for the Blue Nile. Water Resources Res. 2014;50:2086–107.

Kaiser B. Water management under constraints: the need for a paradigm shift. In: Kugelman M, Hathaway RM, editors Running on empty: Pakistan's water crisis. Washington, D.C: Woodrow Wilson International Center for Scholars; 2009. PP. 45–63.

Kingsolver B. Water is life. National Geographic. 2010;217:236.

Knutson TR, Zhang R, Horowitz LW. Prospects for a prolonged slowdown in global warming in the early 21st century. Nature Communications. 2016;7.

Kovats RS, Valentini R, Bouwer LM, Georgopoulou E, Jacob D, et al. Europe. In: Barros VR, Field CB, editors. Climate change: impacts, adaptation, and vulnerability. Part B: Regional aspects. Contribution of working group II to the fifth assessment report of the inter governmental panel on climate change. Cambridge University Press, Cambridge, United Kingdom and New York, NY, USA, 2014;1267–1326.

Lacis AA, Schmidt GA, Rind D, Rued RA. Atmospheric CO_2: principal control knob governing Earth's temperature. Science. 2010;330:356–9.

Lamarque JF, Dentener F, McConnell J, Ro CU, Shaw M, Vet R, et al. Multi-model mean nitrogen and sulfur deposition from the atmospheric chemistry and climate model inter comparison project (ACCMIP): evaluation historical and projected changes. 2013.

Larsen TA, Alder AC, Eggen RI, Maurer M, Lienert J. Source separation: will we see a paradigm shift in wastewater handling? Environ Sci Technol. 2009;43:6121–5.

Lee S, Speight JG, Loyalka SK, eds. Handbook of alternative fuel technologies.CRC Press; 2014.

Lim YK, Phang SM, Rahman NA, Sturges WT, Malin G. Halocarbon emissions from marine phytoplankton and climate change. Int J Environ Sci Technol. 2017;1–16.

Line DE, GD, Jennings RA, McLaughlin DL, Osmond WA, et al. Nonpoint sources. Water Environ Res. 1999;71:1054–69.

Majewski MS, Capel PD. Pesticides in the atmosphere: distribution, trends, and governing factors. In: Gilliom RJ, editor. Pesticides in the hydrologic system; 1995.

McDonald RI, Green P, Balk D, Fekete BM, et al. Urban growth, climate change, and freshwater availability. Proc Natl Acad Sci. 2011;108:6312–7.

Mckinney MA, Pedro S, Dietz R, Sonne C, Fisk AT, Roy D, et al. A review of ecological impacts of global climate change on persistent organic pollutant and mercury pathways and exposures in arctic marine ecosystems. Current Zool. 2015;61:617–28.

Michael K. Introduction. In: Kugelman M, Hathaway RM, editors. Running on empty: Pakistan's water crisis. Washington, D.C: Woodrow Wilson International Center for Scholars; 2009. PP. 5–27.

Mitsch WJ, Bernal B, Nahlik AM, Mander Ü, Zhang L, Anderson CJ, et al. Wetlands, carbon, and climate change. Landscape Ecol. 2013;28:583–97.

Moldan B, Billharz S, Matravers R. Sustainability indicators. A Report on the Project on Indicators of Sustainable Development Scope. 1997;58.

Palmer TN. Towards the probabilistic Earth-system simulator: a vision for the future of climate and weather prediction. Quart J Royal Meteorol Soc. 2012;138:841–61.

Peters NE, Meybeck M. Water quality degradation effects on freshwater availability: impacts of human activities. Water Int. 2000;25:185–93.

Rahaman MM. Water wars in 21st century: speculation or reality? Int J Sust Soc. 2012;4:3–10.

Rhein MA, Rintoul SR, Aoki S, Campos E, Chambers D, Feely RA, Gulev S, et al. Observations: ocean. Climate Change. 2013;255–315.

Richardson SD, Ternes TA. Water analysis: emerging contaminants and current issues. Anal Chem. 2011;83:4614–48.

Rodhe H. A comparison of the contribution of various gases to the greenhouse effect. Science. 1990;248:1217.

Rosenzweig C, Strzepek KM, Major DC, Iglesias A, Yates DN, McCluskey A, Hillel D. Water resources for agriculture in a changing climate: international case studies. Glob Environ Change. 2004;14:345–60.

Sahin O, Stewart RA, Helfer F. Bridging the water supply-demand gap in Australia: a desalination case study. In: European Water Resources Association (EWRA) 8th International Conference 2013 June. PP. 26–29.

Schewe J, Heinkea J, Gertena D, Haddeland I, Arnelld NW, Clark DB, et al. Multimodel assessment of water scarcity under climate change. Proc Natl Acad Sci. 2014;111:3245–50.

Schwarzenbach RP, Egli T, Hofstetter TB, von Gunten U, Wehrli B. Global water pollution and human health. Ann Rev Environ Resour. 2010;35:109–36.

Shahid S. Impact of climate change on irrigation water demand of dry season Boro rice in northwest Bangladesh. Clim Change. 2011;105:433–53.

Simi K. Pakistan's water challenges: entitlement, access, efficiency, and equity. In: Kugelman M, Hathaway RM, editors. Running on empty: Pakistan's water crisis. Washington, D.C: Woodrow Wilson International Center for Scholars, 2009;28–44.

Stocker, T. ed. 2014. Climate change 2013: the physical science basis: Working Group I contribution to the Fifth assessment report of the Intergovernmental Panel on Climate Change. Cambridge University Press.

Swimme B, Berry T. The universe story. Arkana; 1994.

United Nations, Department of Economic and Social Affairs, World Population Prospects, the 2012 Revision. World Population Prospects, 2012.

Vörösmarty CJ, Green P, Salisbury J, Lammers RB. Global water resources: vulnerability from climate change and population growth. Science. 2000;289:284–8.

World Development Indicators. The World Bank, 18 Dec. 2013. http://databank.worldbank.org/data/databases.aspx.

Xiao-jun W, Jian-yun Z, Shamsuddin S, Rui-min H, Xing-hui X, Xin-li M. Potential impact of climate change on future water demand in Yulin city, Northwest China. Mitigat Adapt Strategies Glob Change. 2015;20:1–9.

Zaman M, Rahman A, Haddad K. Regional flood frequency analysis in arid regions: a case study for Australia. J Hydrol. 2012;475:74–83.

Zimmerman J B, Mihelcic JR, Smith AJ. Global stressors on water quality and quantity. 2008;4247–54.

Chapter 3
Water Sustainability Issues in Biofuel Production

Muhammad Arshad and Mazhar Abbas

Abstract Biofuel production process use fresh water collected mainly from surface water flows or from underground natural reservoirs for different activities and it became contaminated with organic and inorganic pollutants. Waste water quality returned to soil and to surface water flows is very poor. To produce one liter of ethanol, 10–17 L of water are consumed. Biofuel production plants are water intensive and there is an upward trend in water consumption. The chapter will describe agricultural and industrial activities involving current water consumption during biofuel production. Major steps of lifecycles for biofuel production pathways: bioethanol from sugarcane molasses and cellulosic feedstock, Biogas from distillery spent wash and Biodiesel from various sources will be evaluated regarding water consumption. The amount of irrigation water used in growth of biofuel feedstock and water consumption for biofuel production through various processing technologies will be analyzed. The vital importance of water management during the feedstock production and conversion stage of the biofuel's lifecycle will also be discussed.

Keywords Climate change · Water sustainability · Biofuel · Fresh water

3.1 Introduction

Worldwide governmental policies to come across the unreliable fossil fuel prices, consistency of energy supply, and changes in climatic situation, have increased the share of alternate biofuel in the energy recipe since last decade (Bot et al. 2015). The production of biofuel also carries on expanding due to their environmental and social implications (Yue et al. 2014) globally as the alternate fuel sources to replace conventional gasoline and diesel in transport sector.

M. Arshad (✉) · M. Abbas
Jhang-Campus, University of Veterinary and Animal Sciences, Lahore, Pakistan
e-mail: Muhammad.arshad@uvas.edu.pk

M. Abbas
e-mail: Mazhar.abbas@uvas.edu.pk

M. Arshad (ed.), *Perspectives on Water Usage for Biofuels Production*,
https://doi.org/10.1007/978-3-319-66408-8_3

Currently, most outstanding issue of the society is to carrying out proper strategies for the mitigation of climate change with particular attention to greenhouse gas emissions (GHG) (Duguma et al. 2014). The transport sector is responsible for almost 19% of the total emissions in Europe (EU27) (Quadrelli and Peterson 2007) and the use of biofuel is strongly recommended. For now, the EU27 member states have set their state marks for biofuel production and heated the debate in scientific community about the potential impacts deriving from a large scale use of biofuel.

The production of biodiesel and bioethanol was 10,505 and 49,540 million (M) liters in 2007 respectively (Gheewala et al. 2013). It became abruptly increased up to 21,463 and 86,986 M liters in 2012, (Earth Policy Institute 2012). Current production of bioethanol is approximately four times higher than the biodiesel, worldwide. Two countries, Brazil and United States of American (US) produce more than 87% of the global ethanol. Many feedstock can be utilized to produce bioethanol, (commercial fuel ethanol) but major feedstock are sugarcane, corn and cassava (Gupta and Verma 2015).

The most obligatory resources for production of biomass are sunlight, macro or micro nutrients, and water (Fageria et al. 2010). Sunlight is almost no limiting and nutrients needed have become a key concern due to price rise of inorganic fertilizers as well as their environmental issues regarding runoff-pollution (Loehr 2012). In conclusion water is specifically significant resource as the race for limited water assets has become more intensive (Gleick 2014). Water is an ever demanded valuable resource, have vital application in human life such as drinking, washing, municipal consumptions, hydropower generation, refrigeration purpose, manufacturing process, recreational activities, habitation for aquaculture and agriculture (Rouse 2016; Loucks et al. 2005). Recent strong stimulus for the huge production of biofuel has affected the availability and quality of water. It is much difficult to monitor the water quality constantly (Jager et al. 2015). Figure 3.1 shows the water

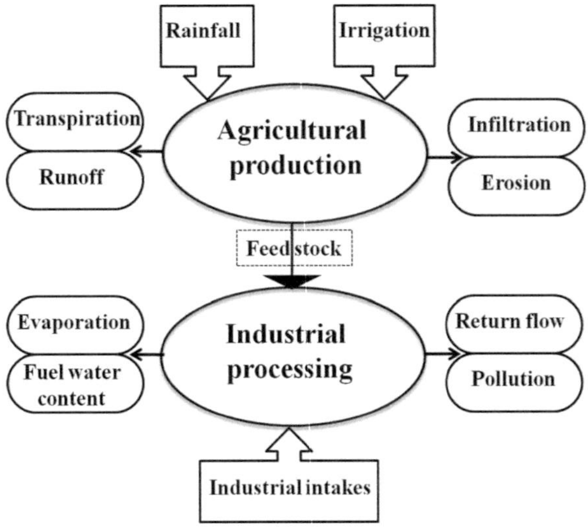

Fig. 3.1 Losses of water in agricultural feedstock production and industrial processing of biofuels

losses ways during production of biofuel. The world has to make decisions on water allocations regarding the production of fuel, food, and fiber.

Although, bio renewable sources are more water-intensive as compare to fossil sources. But least attention has been given to water resources used in production of biofuel and grown-up of their feedstock. (King and Webber 2008; Chiu et al. 2009; Nilsalab et al. 2016). Presently, agricultural sector consumes up most of world water supply (FAO 2008) but cultivation of biomass on huge scale for biofuel production can change the future water demand for agriculture.

Upcoming 50 years will have a 50–70% increased water demand due to growth in world population and nutritional modifications (Jalava et al. 2016). Water is used up throughout the life cycle; starting from feedstock production to its processing into biofuel. Main uses of water are in agricultural farming and bio-refineries. Table 3.1 shows the water use in production of various crops as feedstock of ethanol and biodiesel.

Biofuel are classified into three categories based on their feedstock from which they are produced: first, second, and third-generation. Figure 3.2 shows the sources of bioethanol production from first, second and third generation resources. First-generation biofuel use food crops for production, while second-generation

Table 3.1 Water use in production of various crops as feestock of ethanol and biodiesel (De Fraiture et al. 2008; De Fraiture and Berndes 2009)

Crop	Total WF	Blue WF	Green WF	Total Water	Blue water	Green water
Etahanol		m^3 per GJ ethanol			L of water per of ethanol	
Sugar beet	59	35	24	1388	822	566
Potato	103	46	56	2399	1078	1321
Sugar cane	108	58	49	2516	1364	1152
Maize	110	43	67	2570	1013	1557
Cassava	125	18	107	2926	420	2506
Barley	159	89	70	3727	2083	1644
Rye	171	79	92	3990	1846	2143
Paddy rice	191	70	121	4476	1641	2835
Wheat	211	123	89	4946	2873	2073
Sorghum	419	182	238	9812	4254	5558
Biodiesel		m^3 per GJ biodiesel			L of water per L of biodiesel	
Soybean	394	217	177	13,676	7521	6155
Rap seed	409	245	165	14,201	8487	5714
Jatropha	574	335	239	19,924	11,636	8288

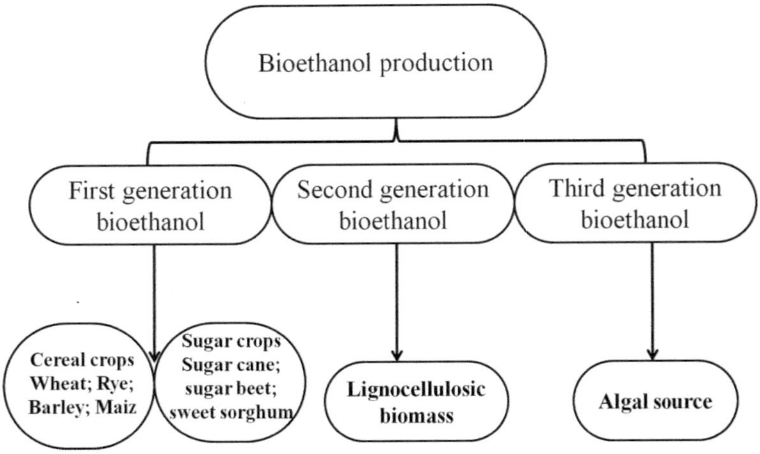

Fig. 3.2 First second and third generation sources of bioethanol production

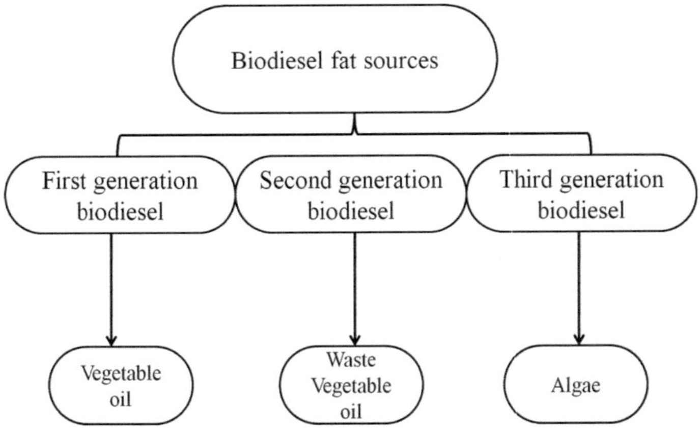

Fig. 3.3 First second and third generation sources of biodiesel production

biofuel' feedstock include woody or fibrous biomass, such as nonedible remains of food crops, or dedicated biofuel crops (Fischer et al. 2009). Third generation biofuel are defined as future biofuel that are derived from microscopic organisms, such as microalgae. First, second and third generation sources of biodiesel production have been shown in Fig. 3.3. At present, most biofuel in commercial use belongs to first generation (Stromberg and Gasparatos 2012; Fischer et al. 2009). Such biofuel have drawbacks, such as small net energy yields, large land, water use, and direct competition with food production (Singh et al. 2011).

Frequent water supply of ample quality is the essential to sustain nature and to feed the increasing human worldwide. Availability of fresh quality water shows a discrepancy with time and space. Requirements of additional food joint with shift from fossil energy towards bioenergy needs an increased demand of water (UNEP 2009). In this chapter a brief account of water consumption in production of bioethanol, biodiesel and biogas is presented.

3.2 Water Demand for Oil Crop Based Biodiesel

During formulation of plans and policies regarding development of biofuel due to environmental concerns, major focus remained on the reduction of GHG emissions (Babel et al. 2011; Pimentel et al. 2008; Schoneveld et al. 2011; Jindal and Goyal 2012) rather to put emphasis on water sources. Recent extra requirements of water to attain the biofuel production goals will put additional burden on water resources. Production of biofuel requires additional water as compared to crude oil. It ranges between 24 and 143 m^3/GJ for the production of biofuel (Okadera and Chontanawat 2010).

Worldwide production of biodiesel has gained pace in recent years. Many countries have set their targets globally for the production of biofuel. Various crops like oil palm, soybean and canola are being utilized as feedstock for biodiesel production. Soybean with canola and waste cooking oils are used as feedstock for biodiesel production in US. But soybean is leading (Schill 2008) with almost 17% of entire soybean grown were utilized in methyl esterification process. Therefor spreading out of biodiesel industry, ignite the food versus fuel debate, changes of land use and amplified requirements of water sources. The following concerns may affect the sustainable biodiesel production at industrial scale.

3.2.1 Water Requirement of Soybean Based Biodiesel Production

Tu et al. (2016) studied the water consumption of water during three stages: soybean growth, soybean processing to soybean oil, and biodiesel manufacturing. Water requirements at irrigation, oil processing, and biodiesel production stages were 61.78, 0.17, and 0.31 gal/gal respectively. It indicates that on average irrigation water consumption accounts for 61.78 gal/gal while water consumption in soybean oil processing and biodiesel manufacturing is 0.17 and 0.31 gal/gal, respectively.

3.2.2 Water Requirement of Oil Palm-Based Biodiesel Production

Nilsalab et al. (2016) estimated the water requirement for oil palm crops in Thailand. According to them 0.33 million m^3ha^{-1} water is required to grow the oil palm trees. The amount of additional freshwater withdrawal in the three regions during the dry season is higher than that during the wet season for both starting and established periods, which is of course reasonable. The largest amount of additional freshwater withdrawal for both periods in the dry season as compared to the wet season is found in the central region. In case of fulfilling the total crop water requirement in certain periods of time that have relatively less rainfall, the total amount of additional freshwater withdrawal in the range of 5311–9591 $m^3ha^{-1}year^{-1}$ would probably be required from freshwater sources. This is quite high compared to the demands of cassava and sugarcane, which may need around 1239–2152 and 1992–3480 $m^3ha^{-1}year^{-1}$ of additional freshwater withdrawal or non-biofuel crops, e.g., maize and soybean, which averages of 2741 and 2276 $m^3ha^{-1}year^{-1}$ of additional freshwater withdrawal (Gheewala et al. 2014).

3.2.3 Water Required for Processing of Feedstock into Biodiesel

Refining of crude palm oil extraction and the production of biodiesel require water, majorly associated use in boilers. Quantity of water used these processes amounts to 0.0022, 0.0001, and 0.00177 m^3L^{-1} of biodiesel, in that order. The amount of water used in the processes of biodiesel production is much less compared to that required for the cultivation of oil palm, contributing only around 0.1% of the total. So, the average water requirement for growing oil palm is much larger than the water requirement for producing biodiesel. Additionally, the efficiency of water used in plantation is also based on how well farmers meet the standard of good agricultural practices to retain the peak crop productivity during that age (Basiron 2007; Silalertruksa and Gheewala 2012; Arshad et al. 2014a).

3.3 Microalgae Based Biodiesel

Microalgae received ample attention as a promising feedstock and will be the huge source of liquid biofuel for transportation in 2030 (European Commission 2010). Algae are single cell microbes with photosynthesis capability, growing in aqueous environment. A number of algal strains keep lipid contents above 70% on dry weight basis (Menetrez 2012). Algal species can be categorized into: freshwater algae and saline algae. Different types of biofuel can be got from microalgae, depending upon the composition of the biomass. Figure 3.4 shows the processes flow diagram of biofuel production from algal sources.

Increasing interest in algal biofuel needs attention regarding their water requirements. As water shortage is a rising concern, therefor well-organized utilization of this limited source should be an important characteristic for biofuel crops. The work of some earlier authors on water use in production of various crops as feestock of ethanol and biodiesel is presented in Table 3.1. They have tried to evaluate the water utilization of microalgae biofuel but with limited information. The water footprint is one of the basic aspects that may restrict the scaling up of the production of biofuel from microalgae at industrial level, as freshwater is shortened in many parts of the world.

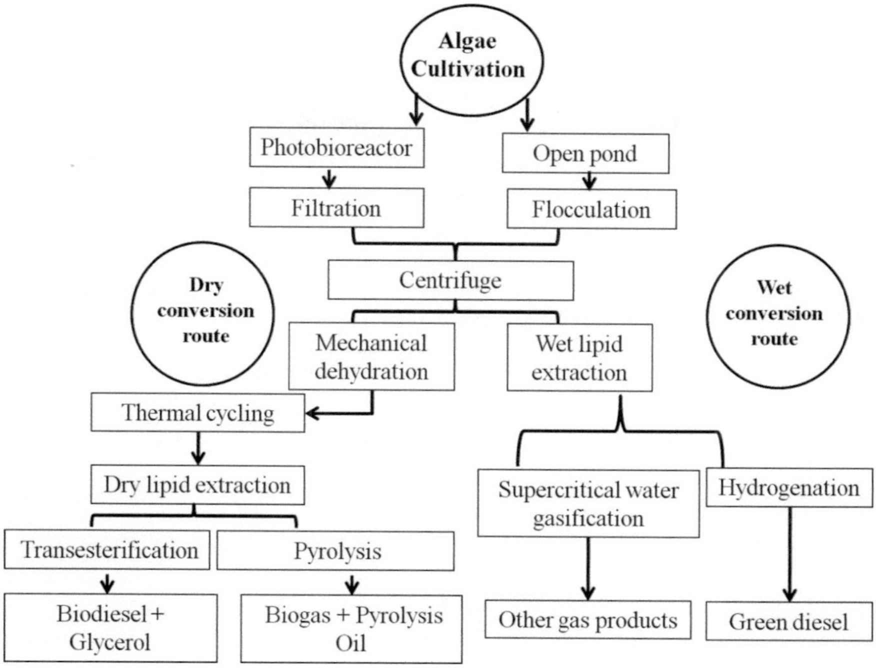

Fig. 3.4 Algal cultivation for the production of biofuels (Ugwu et al. 2008; Ozkan et al. 2012)

There are five major steps, cultivation, harvesting, dewatering, oil extraction and oil upgrading in entire chain of biofuel production from microalgal biomass. The algal microbes need water for their growth can be classified into two types of cultivation systems: open pond systems and closed photobioreactors (PBRs). Table 3.2 shows the comparison of algal cultivation in open pond and photobioreactor systems.

Table 3.2 Comparison of algal cultivation in open pound and photo bioreactor

Variable	Open ponds	Photobioreactors	References
Costs	Cheap	Expensive	Brennan and Owende (2010)
Land use	High	Low	Chisti (2007)
Growing environment	Well controlled	Poorly controlled	
Productivity	Low	High	
Oil yield	35–45 $m^3 ha^{-1} yr^{-1}$	50–60 $m^3 ha^{-1} yr^{-1}$	
Biomass concentration	0.14 kgm^{-3}	4.00 kgm^{-3}	
Water evaporation	High	Low	Brennan and Owende (2010)

3.3.1 Collection of Algal Biomass

The micro algal slurry keeps water up to almost 99.9% of the fresh weight collected from open ponds. It needs to be dried to a definite dryness level. The techniques used are flocculation, gravity sedimentation, ultrasonic aggregation and flotation to attain certain solid matter (Brennan and Owende 2010). The microalgae slurry that is water suspended contains approximately 0.05% dry weight of microalgae (Xu et al. 2011). PBRs give rise to slurry of 2% dry weight after filtration. Thermal techniques can make algae slurry to a higher solid matter concentration of at least 85%.

3.3.2 Water Consumption for Production of Algae Biomass

The biofuel presently in use are found to reduce amount of greenhouse gases but more water exhaustive as compared to fossil fuels (Chiu et al. 2009; Mekonnen et al. 2011). A great concern on the water supplies for next-generation biofuel is present globally. From third generation biofuel; algal biofuel have been pore over completely regarding environmental effects (Campbell et al. 2011; Vasudevan et al. 2012). Water in production of biofuel from algal biomass has been shown in Fig. 3.5. Utilization of the water in microalgal production for biofuel may become a potential barrier in expansion of this feedstock at industrial scale (Liu et al. 2012; Clarens et al. 2010).

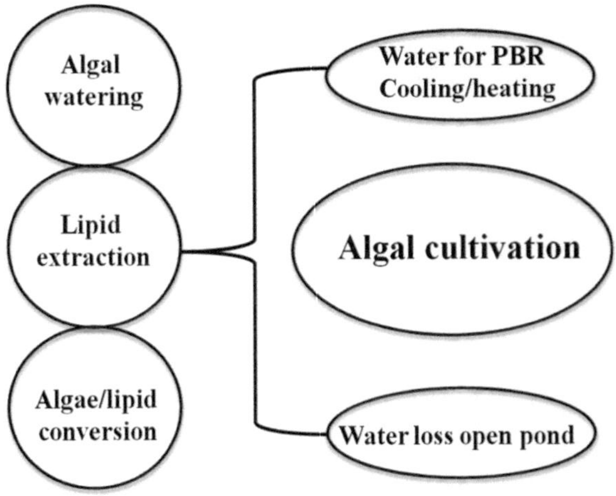

Fig. 3.5 Water in production of biofuels from algal biomass

3.3.3 Technical Routes for Conversion of Algal Biomass into Biofuel

Many technical routes are used to convert microalgal biomass into biofuel but normally following routes are adopted. The thermochemical processes produce oil and gas while biochemical routes produce bioethanol and biodiesel and combustion generates heat and electricity (Amin 2009). To emphasize on water requirement of microalgae biofuel, the conversion routes can be distinguished into wet and dry route technologies.

3.3.3.1 Wet Techniques

Fermentation process. Microbes convert the algal biomass into bioethanol.

Anaerobic digestion. Organic matter is decomposed into biogas consisting of methane (CH_4) and CO_2 in absence of oxygen. Biogas can be straightaway applied for combustion to harvest the energy.

Hydrothermal liquefaction. The algae biomass can be degraded to bio-oil keeping energy stuffing almost equal to biodiesel (US-DOE 2010).

3.3.3.2 Dry Route Technologies

Direct combustion. The algal matter having moisture content less than 50% can be applied for combustion in power plants to generate electricity (Brennan and Owende 2010) and produce heat.

Pyrolysis. The technology breaks the algal matter into charcoal, condensable vapors in absence of oxygen.

To determine the water consumed for the production of a product, water footprint (WF) is calculated. Water footprint measures the dimensions of freshwater consumed for the production of a particular product (Hoekstra et al. 2011). It includes the three parts that is green, blue and grey water that differentiate among direct and indirect water consumption. Water footprint for biofuel produced from microalgae has generally two components: grey water and blue water as all nutrients are recycled and grey water part is almost zero.

Quinn (2013) modeled ten locations to study, there annual biomass production ranged from 29.5 to 53 ton $ha^{-1}year^{-1}$, and microalgae oil yields range from 13 to 23.7 $m^3ha^{-1}year$. His results were similar in productivity as measured under large-scale production (Quinn et al. 2011). Overall blue WF remained 42 m^3GJ^{-1} while green WF was negative as water gained in the water basin due to precipitation. Total WF as sum of blue and green WFs, varied amongst the geographies and processes considered between 18 and 82 m^3GJ^{-1}.

All of the water utilized in the processes, water consumed in energy and materials input, and the water credits associated with the coproducts includes in lifecycle WF. It must exclude the water recollected in the water basin. The micro-algal lifecycle WFs differs among geographic and conversion pathways for fuel production. Following precautions must be in consideration while we have to study the waterfoot print of algal biofuel.

1. In case of open pond evaporation for algal biofuel production, longer data series may provide better and dependable results.
2. For the PBRs, there are also some reservations. Evaporation of water PBRs can be reduced in a closed cooling system.
3. Generally, it is assumed the filtration of the first harvest step has no water utilization. But industrial level harvesting by filtration may not be practical. The other methods may be water intensive, involving water loss.
4. There must be no gray WF means the water must be cleaned before discharge.
5. Some PBRs such as modeled by Slegers et al. (2011) for the Netherlands, France, and Algeria, employ both heating and cooling. So the energy required for heating is not taken into account.
6. The energy required for the treatment of wastewater must be considered.
7. Up scaling of the algal systems is associated with, several concerns, such as the CO_2 supply.

3.4 Water Demand for Biogas Production

To analyze, understand and exploration of remedies for environmental crunches, insecurity of food, changing climatic conditions and their interrelationships are the need of the day. Water is a key player to understand the relationship and most of the times, a major limiting factor, with population growth and climate change compromising its availability (Morrison et al. 2010). Estimation of water consumption provides the basics for sustainable development.

Being most propagated source among the biofuel, biogas is a significant to combat environmental impacts, but the struggle in the food market as well as the advances in water uses, dire needs to analyze the nexus between biogas production and water (Stillwell et al. 2010). Biogas produced from anaerobic digestion process can be utilized directly (contaminants like hydrogen sulphide or other sulphur compounds, and particulate matter must be removed) in combustion engines, or may be improved as bio-methane with the elimination of carbon dioxide that can be injected into the natural gas grid or can be utilized as fuel gas (Adelt et al. 2011).

Water resources are limited and being used for growth of energy crops. There, operations of biogas digesters also possibly face difficulty. Therefore environmental performances of biogas production must be analyzed.

3.4.1 Water Requirement of Biogas Crops

Biogas systems based on energy crops need to be analyzed from a comprehensive perspective, considering ecological functions in agricultural and natural landscapes as well as broader livelihood and development implications (Braun et al. 2008). The water necessities of any crop specie need to be analyzed before its inclusion into agricultural system (Ings et al. 2013; Mantovani et al. 2014). For biogas production maize is the most used bioenergy crop in Germany. Out of total 2.49 M ha agricultural area, one third is utilized for cultivation of maize for biogas production (Schoo et al. 2017).

Maize is highly water intensive crop, so there is continuous search for substitutive biogas crops. In this way, the perennial cup plant is receiving increasing attention. It is a second generation bioenergy crop that is not used in food or feed preparations. The environmental effects of pesticides and fertilizers can also be avoided using cup plant as biogas crop (Sanderson and Adler 2008). Cup plant species is always characterized as drought tolerant (Sontheimer 2007; Baubock et al. 2014; Franzaring et al. 2014).

3.4.2 Water Required for Anaerobic Digestion

According to Orskov et al. (2014), 200 dm^3 of water is needed in anaerobic digestion of 10 kg of manure dry matter. Quantity of water required to get optimum biogas production is calculated by the fresh weight of manure produced by livestock and the percentage dry matter in the dung.

Pacetti et al. (2015) evaluated water footprint and environmental performance of biogas production from the anaerobic digestion of energy crops. They have discussed eighteen situations, to check out the performances of different combinations of locations, biogas crops and treatments. The growth phase of the three crops is the one with the most impact; sorghum is the crop with lowest impacts, followed by maize, while wheat has the highest impacts of the three. Biogas production from energy crops has, in general, negative impacts (and so is beneficial), for all the indicators except water depletion, fresh water eco-toxicity and marine eco-toxicity.

3.4.3 Submissions to Improve Water Usage in Biogas Production

Following submissions can improve water use efficiency in biogas production.

Application of wastewater. The wastewater may be used in biogas digester to make efficient reuse of wastewater. It may only meet a proportion of the additional water required for anaerobic digestion. Therefore, before the installation of the digester, methods that will be used to collect the additional water needed for the digester should be considered (Rasi et al. 2011).

Collection of the water needed can take a significant amount of time. Therefore, before installation of the digester, the time spent doing different activities should be budgeted to ensure that the total time spent on household activities does not significantly increase with the installation of the digester.

Application of Rainwater. Rainwater may be harvested for use in digester. Therefore, a possible system with a higher rainwater coefficient should be used to harvest rainwater (Kahinda and Taigbenu 2011). An open pond or ground catchment should be used to collect additional water. Fish or related aquaculture may be maintained in the ponds (Bansal et al. 2016). The alternative biomass sources may be explored to feed the biogas digesters.

3.5 Water Requirements for Production of Bioethanol

Worldwide applications of biofuel especially bio-ethanol takes considerable extra water use for agricultural purposes, that increased competition for water (Fraiture and Berndes 2009). It adds further in decline of water quality due to the seepage of pesticides and fertilizers used (Gallagheret al. 1996; UNEP 2009). Already significant signs of water consumption and pollution above sustainable levels are there, as in the Ganges of India and Indus river basins in Pakistan.

In last three decades, worldwide production of bioethanol has remained greater than ever before. Figure 3.6 shows crop resources used for bioethanol production and also as sweeteners. Brazil and United states (US) are major producers. While US produces it from corn and but Brazil's production is based on sugar cane

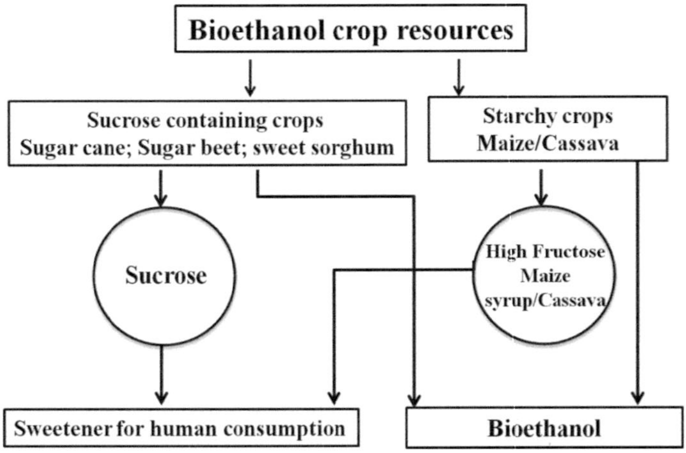

Fig. 3.6 Crop resources used for bioethanol production and also as sweeteners (Gerbens-Leenes and Hoekstra 2012)

(Crago et al. 2010). Transport sector uses about one third of total energy utilized in developed countries, so the shift towards biofuel will take huge efforts.

For production of first generation bioethanol about 60% is shared by sugar crops, majorly sugar cane and less from sugar beet, the remaining from grains, especially from corn (Berg 2004; Arshad et al. 2017). Both sugarcane and sugar beet with maize are valuable food crops with major contribution to worldwide food production (Oerke and Dehne 2004). The sugarcane provides 29%, maize 14% and sugar beet 4% of overall crop production globally. The increased bioethanol production will increase food prices with decrease in food security (FAO 2008; Fischer et al. 2009; Arshad et al. 2014b). Sugarcane like agricultural crops are much water intensive. Presently, agriculture sector uses almost 86% of the worldwide freshwater. For the answer of the question is; can we apply rare water resources for food or fuel. It requires complete information about how much water is required to produce food and fuel.

The water footprint (WF), a tool to calculate water requirements for consumer products is applied by various authors (Hoekstra et al. 2011). Hoekstra and Hung (2002) is premier in preparing first global estimation of freshwater needed to produce crops. Subsequent studies, e.g. for cotton and coffee and tea provide detailed WFs of crops and derived crop products. Sugar crops are among the most important food crops that, at the same time, are used to produce growing amounts of bio-ethanol.

Several studies assessed bio-ethanol water requirements. Chiu et al. (2009) and King and Webber (2008) made assessments for US bio-ethanol from maize focusing on irrigation water.

3.5.1 Water Requirements for Bioethanol Crops Cultivation

Globally fresh water is majorly used in conventional agriculture to raise food and fiber. According to FAO (2008) most of freshwater taking out from rivers, lakes, and aquifers are used in agriculture worldwide. Since last decade, fossil fuel prices are on the rise with the harmful impact on the environment through carbon dioxide emissions has compelled many nations to explore for various biofuel crops to produce ethanol, biodiesel and biogas for fuel (US-DOE 2008). Brazil and United States are leaders in bioethanol production, production it from two major crops, sugar cane and corn. The crops have been analyzed and compared with respect their relative water utilization during the production of the biofuel feedstock.

3.5.1.1 Sugarcane Production in Brazil

Sugarcane is the tall annual crop innate to warm temperate and tropical areas. Its fibrous stalks are full of sugar and sized 2–4 m in length. Ready availability of sugars in higher concentration makes it matchlessly fit for bioethanol production through microbial fermentation. Brazil is renowned as the world's leader in production of bioethanol (DOE-EIA 2007; Trostle 2008) from sugarcane. The country wide promotion of the sugarcane crop for ethanol production was started from 1970 after the first worldwide fuel crisis (Rother 2006). Brazil is the global leader in production of sugarcane.

The utilization of water in whole ethanol production process can be regarded as crop production and ethanol production. Normal 8–12 mm water is used to raise approximately one ton of sugarcane. As the sugarcane grows throughout the year, its annual requirements for water are about 1500–2500 mm/year (Goldemberg et al. 2008; Moreira 2007). Rainfall with partial irrigation is the major source of water of the sugar cane plantation in Brazil. So the irrigation water use is generally minor. But the increasing demand of ethanol, expanded the sugar cane production and need to complement rainfall with irrigation is on rise (Goldemberg et al. 2008).

Conversions of sugarcane to bioethanol utilize huge quantity of water. Majorly water is required in four key processes: cane washing; concentration of juice extracted from sugarcane through evaporation; in fermentation process for dilution of substrate and cooling of fermenters; in alcohol distillation for condenser's cooling (Arshad et al. 2008, 2011; Arshad 2010). Almost 21 m^3/ton of sugar cane is utilized. However, recycling of the waste water can reduce the fresh water consumption (Macedo 2005). According to Goldemberg et al. (2008), the efficient production processes can reduce the consumption of water from 5.3 to 1.83 m^3/ton of sugarcane.

3.5.1.2 Corn Production in US

Currently, corn is major feedstock for ethanol production, likely to be increased as the renewable biofuel are thirstily required. It is estimated that almost one fourth of the US domestic corn used for ethanol production. Mostly corn producing states have sufficient rainfall with adequate water holding capacity of soil to raise corn without additional irrigation. In some corn producing states irrigation is needed to grow corn and major source for irrigation water aquifer (Maupin and Barber 2005). The said Ogallala Aquifer is also provides water for municipal and industrial users. Incidence of frequent droughts, with the irrigation systems has lowered the water levels in the aquifer of most regions. The water present in the aquifer is normally called as geologic water as it is a general concept that rainfall takes many years to reach aquifer (Andrews et al. 1999). Such slow water recharge coupled with huge utilization of water results lowering water tables over most of the aquifer.

3.5.1.3 Comparison of Sugar Cane and Corn Water Usage

Sugar cane and corn is photo synthetically very different to each other. Much water is needed to raise corn grain as compared to sugarcane. Key reason behind the extra water use for corn grain is that only grain is presently used in bioethanol

Table 3.3 Water requirements comparison between crops used for biofuel production (Stone et al. 2010)

Crops	Water requirement (m^3water/Mg crop)	Biofuel conversion (L fuel/Mg crop)	Crop water requirement for biofuel (m^3 water/Mg fuel)	Crop water requirement per unit energy (m^3 water/GJ)
Ethanol				
World corn (grain)	833	709	2580	97
World sugarcane	154	334	580	22
Nebraska corn (grain)	634	409	1968	74
Corn Stover	634	326	2465	92
Corn Stover + Grain	634	735	1093	41
Switchgrass	525	336	1980	74
Grain Sorghum	2672	358	9460	354
Sweet Sorghum	175	238	931	35
Biodiesel				
Soybean	1818	211	9791	259
Canola	1798	415	4923	13

production. Water requirements comparison between crops used for biofuel production is presented in Table 3.3. With the advent new technologies for conversion of cellulose feedstock into bioethanol production, comparative difference between the crop water needs will be minimized. But the use of corn biomass for ethanol production may reduce the recycling of minerals and nutrients to soil (Mann et al. 2002; Doran et al. 1984; Wilhelm et al. 2004). Moreover the shorter growing season of corn with higher water demands can make corn grain production vulnerable to short term droughts and lack of supplemental water supplies for irrigation (Eaves and Eaves 2007).

3.5.1.4 Sweet Sorghum

Sweet sorghum has been well known as a potential biofuel crop for ethanol production worldwide. It is drought-resistant bioenergy crop and can be broadly adjusted in several growing conditions. Mengistu et al. (2016) has studied the water use efficiency of drip-irrigated sweet sorghum under two different climatic conditions in South Africa. Seasonal water use was estimated using eddy covariance and surface renewal methods. Bioethanol water usage efficiency for the sweet sorghum at Ukulinga were 0.27 and 0.60 L m^{-3} for 2010/11 and 2011/12 growing seasons, respectively. The WUE of sweet sorghum was sensitive to plant density.

3.5.1.5 Cassava

Cassava is possibly amongst the best crops in terms of carbohydrate production (Wang 2002; Bokanga 1996) and grows in areas where temperature ranges from 291.15 to 298.15 K. For its better growth, 0.05–5 m rainfall annually is needed (Adeoti 2010). This root crop is highly rich in carbohydrate (Ziska et al. 2009) and the highest producer of carbohydrates after sugarcane.

Adeoti (2010) has evaluated the virtual water content of fresh cassava root and water needed during the entire period of cassava crop growth. He estimated that water footprint for fuel ethanol in is averaged at 15.3 m^3kg^{-1} from cassava. Production of ethanol will need 6.0 km^3y^{-1} of water, or about 3% of the estimated water resource of Nigeria.

Cultivation time is generally different in various regions; the sugarcane and Cassava are normally harvested within 10–12 months after plantation, although their cultivation time is different (Suksri et al. 2007).

3.5.2 Water Requirements for Bioethanol Feedstock Processing

For the case of molasses based ethanol production, the sugar mill is involved as the feedstock processing step to produce molasses feedstock. Sugar milling involves crushing cane to extract sugarcane juice. This juice is clarified to remove any impurities and concentrated into syrup by boiling off excess water, seeded with raw sugar crystals in a vacuum pan and boiled until sugar crystals have formed and grown (Silalertruksa and Gheewala 2009). The crystals are separated from the syrup by centrifugal process before more crystals are grown in the syrup. Therefore, a variety of products and wastes will be generated in the mills i.e. sugar is the main product; molasses, the syrup remaining after the sugar has passed through the centrifuge for the last time in a mill or refinery, is a byproduct as well as bagasse which is generated after sugarcane crushing and it used to produce steam and electricity to supply for the mills and the surplus electricity is sold to the general grid-mix.

The other residues such as filter cake and wastewater effluents from the mills are considered as waste in the study because generally they done not have an economic value and are hence, not traded. To share the water use from sugarcane cultivation and sugar milling between the sugar (main product) and the by-products i.e. molasses and bagasse, the energy-based allocation techniques is applied in the study. In the mills, a ton of sugarcane processed will generate 109, 45, and 287 kg of sugar, molasses, and bagasse, respectively (Arshad and Ahmed 2016). However, only the surplus bagasse after internal use in the mills (for own energy requirements) i.e. about 131 kg per ton sugarcane, that will be considered in the allocation calculation.

Based on the average energy content of sugar, molasses and bagasse of about 16.33, 11.43 and 7.53 MJ/kg respectively (Silalertruksa and Gheewala 2009), the allocation factors for sugar, molasses and the surplus bagasse are 0.54, 0.16 and 0.30, respectively. The factor of 0.16 is used for determining the water use for molasses production. Based on all processes including sugarcane washing, extraction, juice treatment, juice concentration by condenser and evaporation (excluding ethanol production), the water use is estimated to be around 1.23 m^3 per ton of processed cane (Macedo 2005).

The water use in sugar mills for molasses is estimated to be around 4.37 m^3/ton molasses. The water use in the industrial processes i.e. feedstock processing and ethanol conversion are considered to contribute to the blue WF. Effluents generated in this process contribute to water pollution. As cassava ethanol is mainly produced from fresh cassava root and also the dried chip processing step does not require the water; therefore, the process of converting fresh cassava to dried chips form is not account in the study.

3.5.2.1 Water Requirements for Bioethanol Conversion

There are two routes of ethanol production from sugarcane feedstock. It is produced from sugarcane molasses, the byproduct of sugarcane milling and can be produced directly from the sugarcane juice.

Molasses ethanol. The processes of molasses ethanol production consist of yeast preparation, fermentation, distillation and dehydration. The study refers production data of molasses ethanol conversion from literature (Silalertruksa and Gheewala 2009). To produce a liter (L) of molasses ethanol, around 4.6 kg of molasses is required or equivalent to around 61 L of molasses ethanol/ton of sugarcane. The water used at the ethanol conversion stage is about 8.6 L/L of molasses ethanol. It is classified as "blue water" and the two main water-intensive processes are the fermentation and the supporting process as steam generation.

Sugarcane ethanol. To produce sugarcane ethanol, sugarcane juice which is extracted from sugarcane crushing process will be directly used to produce ethanol without the production of sugar. Bagasse is used to produce steam and electricity. Blue water required at this stage is about 14.3 L/L sugarcane ethanol and it is mainly for steam production. Spent wash is sent to aerobic ponds with biogas recovery system. To produce a liter of sugarcane ethanol, around 11.6 kg of sugarcane is required or equivalent to around 86 L of sugarcane ethanol/ton of sugarcane (Silalertruksa and Gheewala 2011).

Cassava ethanol. The cassava ethanol plant consists of five main processes i.e. (1) cassava preparation including cleaning.

3.6 Conclusion

It is need of the day, to evaluate and understand the water sustainability of biofuel, as many countries have increased their production mandates. Water sustainability must be a significant factor before consider the future investments in biofuels. Use of water in production of feedstock is vital contributor to environment and climatic changes impact of biofuels. Efficient use of land and water resources is prime necessity now. According to current water situation in many regions of the world, especially in developing countries, some key questions have to be answer before consider the future biofuel targets:

1. Is there enough water present to meet our needs?
2. In the foreseeable future and in the distant future, enough water will be available? Are the water resources limited?
3. How does biofuel and bioenergy production influence water resources and vice versa? Is there enough water present to meet increasing biofuel needs?

How much socioeconomic development and environmental management will be affected, when water resources will be directed towards biofuel production?

References

Adelt M, Wolf D, Vogel A. LCA of biomethane. J Nat Gas Sci Eng. 2011;3:646–50.

Adeoti O. Water use impact of ethanol at a gasoline substitution ratio of 5% from cassava in Nigeria. Biomass Bioenerg. 2010;34:985–92.

Amin S. Review on biofuel oil and gas production processes from microalgae. Energ Convers Manage. 2009;50:1834–40.

Andrews WJ, Osborn NI, Luckey RR. Rapid recharge of parts of the high plains aquifer indicated by a reconnaissance study in oklahoma. US Geological Survey Fact Sheet 137-00; 1999. http://www.owrb.ok.gov/studies/reports/reports_pdf/high_plains_2.pdf.

Arshad M. Bioethanol: A sustainable and environment friendly solution for Pakistan. A Sci J COMSATS–Sci. Vision. 2010;16–7.

Arshad M, Ahmed S. Cogeneration through bagasse: a renewable strategy to meet the future energy needs. Renew Sust Energ Rev. 2016;54:732–7.

Arshad M, Khan ZM, Shah FA, Rajoka MI. Optimization of process variables for minimization of byproduct formation during fermentation of blackstrap molasses to ethanol at industrial scale. Lett Appl Microbiol. 2008;47:410–4.

Arshad M, Zia MA, Asghar M, Bhatti H. Improving bio-ethanol yield: Using virginiamycin and sodium flouride at a Pakistani distillery. Afr J Biotechnol. 2011;10:11071.

Arshad M, Adil M, Sikandar A, Hussain T. Exploitation of meat industry by-products for biodiesel production: Pakistan's perspective. Pakistan J Life Soc Sci. 2014a;12:120–5.

Arshad M, Ahmed S, Zia MA, Rajoka MI. Kinetics and thermodynamics of ethanol production by *Saccharomyces cerevisiae* MLD10 using molasses. Appl Biochem Biotechnol. 2014b;172:2455–64.

Arshad M, Hussain T, Iqbal M, Abbas M. Enhanced ethanol production at commercial scale from molasses using high gravity technology by mutant *S. cerevisiae*. Brazilian J Microbiol. 2017. http://doi.org/10.1016/j.bjm.2017.02.003.

Babel MS, Shrestha B, Perret SR. Hydrological impact of biofuel production: a case study of the Khlong Phlo Watershed in Thailand. Agric Water Manage. 2011;101:8–26.

Bansal V, Tumwesige V, Smith JU. Water for small-scale biogas digesters in Sub-Saharan Africa. GCB Bioenerg. 2016.

Basiron Y. Palm oil production through sustainable plantations. Eur J Lipid Sci Tech. 2007;109:289–95.

Baubock R, Karpenstein-Machan M, Kappas M. Computing the biomass potentials for maize and two alternative energy crops, triticale and cup plant (*Silphium perfoliatum* L.), with the crop model BioSTAR in the region of Hannover (Germany). Environ Sci Eur. 2014;26:19.

Berg C, Licht FO. World fuel ethanol. Analysis and outlook, report for FO Licht. 2004.

Bokanga M. Biotechnology and cassava processing in Africa. IITA Res. 1996;12:14–8.

Bot P, van Donk DP, Pennink B, Simatupang TM. Uncertainties in the bidirectional biodiesel supply chain. J Clean Prod. 2015;95:174–83.

Braun R, Weiland P, Wellinger A. Biogas from energy crop digestion. InIEA bioenergy task 2008 (vol. 37, pp. 1–20).

Brennan L, Owende P. Biofuels from microalgae-a review of technologies for production, processing, and extractions of biofuels and co-products. Renew Sust Energ Rev. 2010;14:557–77.

Campbell PK, Beer T, Batten D. Life cycle assessment of biodiesel production from microalgae in ponds. Bioresour Technol. 2011;102:50–6.

Chisti Y. Biodiesel from microalgae. Biotechnol Adv. 2007;25:294–306.

Chiu YW, Walseth B, Suh S. Water embodied in bioethanol in the United States. Environ Sci Technol. 2009;43:2688–92.

Clarens AF, Resurreccion EP, White MA, Colosi LM. Environmental life cycle comparison of algae to other bioenergy feedstocks. Environ Sci Technol. 2010;44:1813–9.

Crago CL, Khanna M, Barton J, Giuliani E, Amaral W. Competitiveness of Brazilian sugarcane ethanol compared to US corn ethanol. Energy Policy. 2010;38:7404–15.

De Fraiture C, Giordano M, Liao Y. Biofuels and implications for agricultural water use: blue impacts of green energy. Water Policy. 2008;10(S1):67–81.

De Fraiture C, Berndes G. Biofuels and water. RW Howarth and S. Bringezu (eds.). 2009:139–53.

DOE-EIA, 2007. Country analysis briefs: Brazil http://www.eia.doe.gov/emeu/cabs/Brazil/pdf.pdf . Department of Energy, Energy Information Administration, Washington, DC.

Doran JW, Wilhelm WW, Power JF. Crop residue removal and soil productivity with no-till corn, sorghum, and soybean. Soil Sci Soc Am J. 1984;48:640–5.

Duguma LA, Minang PA, van Noordwijk M. Climate change mitigation and adaptation in the land use sector: from complementarity to synergy. Environ Manage. 2014;54:420–32.

Earth Policy Institute, 2012. Full planet, empty plates Chapter 4 Data: Food or Fuel? http://www.earth-policy.org/?/data_center/C24/.

Eaves J, Eaves S. Renewable corn-ethanol and energy security. Energ Policy. 2007;35:5958–63.

European Commission, 2010. Trends to 2030 update. http://ec.europa.eu/energy/observatory/trends_2030/doc/trends_to_2030_update_2009.pdf.

Fageria NK, Baligar VC, Jones CA. Growth and mineral nutrition of field crops. CRC Press; 2010.

FAO, 2008. Major Food and agricultural commodities and producers: countries by commodity. http://www.fao.org/es/ess/top/commodity.html?lang=en&item=156&year=2005.

Fischer G, Hizsnyik E, Prieler S, Shah M, Van Velthuizen H. Biofuels and food security. Laxenburg, Austria: Int Inst Appl Syst Anal; 2009.

Franzaring J, Schmid I, B€auerle L, Gensheimer G, Fangmeier A. Investigations on plant functional traits, epidermal structures and the ecophysiology of the novel bioenergy species *Sida hermaphrodita* Rusby and *Silphium perfoliatum* L. J Appl Bot Food Qual. 2014;87:36–45.

Gallagher DL, Dietrich AM, Reay WG, Hayes MC, Simmons GM. Ground water discharge of agricultural pesticides and nutrients to estuarine surface water. Ground Water Monit Remed. 1996;16:118–29.

Gerbens-Leenes W, Hoekstra AY. The water footprint of sweeteners and bio-ethanol. Environ Int. 2012;40:202–11.

Gheewala SH, Silalertruksa T, Nilsalab P, Mungkung R, Perret SR, Chaiyawannakarn N. Implications of the biofuels policy mandate in Thailand on water: the case of bioethanol. Bioresour Technol. 2013;150:457–65.

Gheewala SH, Silalertusksa T, Nilsalab P, Mungkung R, Perret SR, Chaiyawannakarn N. Water footprint and impact of water consumption for food, feed, fuel crops production in Thailand. Water. 2014;6:1698–718.

Gleick PH. The world's water volume 8: The biennial report on freshwater resources (vol. 8). Island Press; 2014.

Goldemberg J, Coelho ST, Guardabassi P. The sustainability of ethanol production from sugarcane. Energ Policy. 2008;36:2086–97.

Gupta A, Verma JP. Sustainable bio-ethanol production from agro-residues: a review. Renew Sust Energ Rev. 2015;41:5–567.

Hoekstra AY, Hung PQ. Virtual water trade. A quantification of virtual water flows between nations in relation to international crop trade. Value of water research report series. 2002;11:166.

Hoekstra AY, Chapagain AK, Aldaya MM, Mekonnen MM. The water footprint assessment manual: setting the global standard. Routledge; 2011.

Ings J, Mur LA, Robson PR, Bosch M. Physiological and growth responses to water deficit in the bioenergy crop *Miscanthus x giganteus*. Front Plant Sci. 2013;4:468.

Jager HI, Baskaran LM, Schweizer PE, Turhollow AF, Brandt CC, Srinivasan R. Forecasting changes in water quality in rivers associated with growing biofuels in the Arkansas-White-Red river drainage, USA. Gcb Bioenerg. 2015;7:774–84.

Jalava M, Guillaume JH, Kummu M, Porkka M, Siebert S, Varis O. Diet change and food loss reduction: what is their combined impact on global water use and scarcity? Earth's Future. 2016;4:62–78.

Jindal S, Goyal K. Evaluation of performance and emissions of *Hibiscus cannabinus* (Ambadi) seed oil biodiesel. Clean Technol Environ Policy. 2012;14:633–9.

Kahinda JM, Taigbenu AE. Rainwater harvesting in South Africa: challenges and opportunities. Physics and Chemistry of the Earth, Parts A/B/C. 2011;36:968–76.

King CW, Webber ME. Water intensity of transportation. Environ Sci Technol. 2008;42:7866–72.

Liu X, Clarens AF, Colosi LM. Algae biodiesel has potential despite inconclusive results to date. Bioresour Technol. 2012;104:803–6.

Loehr R. Agricultural waste management: problems, processes, and approaches. Elsevier. 2012.

Loucks DP, Van Beek E, Stedinger JR, Dijkman JP, Villars MT. Water resources systems planning and management: an introduction to methods, models and applications. Paris: UNESCO; 2005.

Macedo IC. Chapter 5: Impacts on water supply. Sugarcane's energy-12 studies on Brazilian sugarcane agribusiness and its suitability, São Paulo Sugar Cane Agroindustry Union. 2005. http://english.unica.com.br/multimedia/publicacao/. Accessed 18 Sept 2008.

Mann L, Tolbert V, Cushman J. Potential environmental effects of corn (*Zea mays* L.) stover removal with emphasis on soil organic matter and erosion. Agric Ecosyst Environ. 2002;89:149–66.

Mantovani D, Veste M, Gypser S, Halke C, Koning L, et al. Transpiration and biomass production of the bioenergy crop Giant Knotweed Igniscum under various supplies of water and nutrients. J Hydrol Hydromech. 2014;62:316–23.

Maupin MA, Barber NL. Estimated withdrawals from principal aquifers in the United States, 2000. US Geological Survey Circular 1279, US Geological Survey, Reston, 2005; VA, 47pp.

Mekonnen MM, Hoekstra AY. The green, blue and grey water footprint of crops and derived crop products. Hydrol Earth Syst Sci. 2011;15:1577–600.

Menetrez MY. An overview of algae biofuel production potential and environmental impact. Environ Sci Technol. 2012;46:7073–85.

Mengistu MG, Steyn JM, Kunz RP, Doidge I, Hlophe HB, et al. A preliminary investigation of the water use efficiency of sweet sorghum for biofuel in South Africa. Water SA. 2016;42.

Moreira JR. Water use and impacts due ethanol production in Brazil. International conference on linkages in energy and water use in agriculture in developing countries. Organized by IWMI and FAO, ICRISAT, India, January 2007. http://www.iwmi.cgiar.org/EWMA/files/papers/Jose_Moreira.pdf.

Morrison J, Schulte P, Schenck R. Corporate water accounting, methods and tools for measuring water use and its impacts. United Nations Environment Programme: United Nations Global Compact, Pacific Institute; 2010.

Nilsalab P, Gheewala SH, Mungkung R, Perret SR, Silalertruksa T, Bonnet S. Water demand and stress from oil palm-based biodiesel production in Thailand. Int J Life Cycle Ass. 2016;1–12.

Oerke EC, Dehne HW. Safeguarding production-losses in major crops and the role of crop protection. Crop Protect. 2004;23:275–85.

Okadera T, Chontanawat J. Water for bio-energy in Thailand. AS. 2010;44:673–9.

Orskov ER, Anchang KY, Subedi M, Smith J. Overview of holistic application of biogas for small scale farmers in Sub-Saharan Africa. Biomass Bioenerg. 2014;70:4–16.

Ozkan A, Kinney K, Katz L, Berberoglu H. Reduction of water and energy requirement of algae cultivation using an algae biofilm photobioreactor. Bioresource Technol. 2012;114:542–8.

Pacetti T, Lombardi L, Federici G. Water-energy nexus: a case of biogas production from energy crops evaluated by water footprint and LCA methods. J Clean Prod. 2015.

Pimentel D, Marklein A, Toth MA, Karpoff M, Paul GS, et al. Biofuel impacts on world food supply: use of fossil fuel, land and water resources. Energies. 2008;1:41–78.

Quadrelli R, Peterson S. The energy–climate challenge: recent trends in CO_2 emissions from fuel combustion. Energ Policy. 2007;35:5938–52.

Quinn JC. Analysis of water footprint of a photobioreactor microalgae biofuel 1 production system from blue, green and lifecycle perspectives. Mechanical and Aerospace Engineering Faculty Publications. 2013; Paper 32. http://digitalcommons.usu.edu/mae_facpub/32.

Quinn J, de Winter L, Bradley T. Microalgae bulk growth model with application to industrial scale systems. Bioresour Technol. 2011;102:5083–92.

Rasi S, Läntelä J, Rintala J. Trace compounds affecting biogas energy utilisation–a review. Energy Convers Manage. 2011;52:3369–75.

Rother L. With big boost from Sugar Cane, Brazil is satisfying its fuel needs. The New York Times. 10-Apr-2006, natl. ed. http://www.nytimes.com/2006/04/10/world/americas/10brazil.html?_r=1&scp=1&sq=&st=nyt&oref=slogin.

Rouse MJ. Water worldwide-drinking water quality regulation: where are we in a continuing evolution? J Am Water Works Ass. 2016;108:20–4.

Sanderson MA, Adler PR. Perennial forages as second generation bioenergy crops. Int J Mol Sci. 2008;9:768–88.

Schill SR. 2008. Sizing up the soybean market. http://wwwbiodieselmagazine.com/articles/2973/sizing-up-the-soybean-market/.

Schoneveld GC, German LA, Nutakor E. Land-based investments for rural development? A grounded analysis of the local impacts of biofuel feedstock plantations in Ghana. Ecol Soc. 2011;16:10.

Schoo B, Wittich KP, Böttcher U, Kage H, Schittenhelm S. Drought tolerance and water-use efficiency of biogas crops: a comparison of cup plant, maize and lucerne-grass. J Agron Crop Sci. 2017;203:117–30.

Silalertruksa T, Gheewala SH. Long-term bioethanol system and its implications on GHG emissions: a case study of Thailand. Environ Sci Technol 2011;45:4920–8.

Silalertruksa T, Gheewala SH. Food, fuel, and climate change: is palm-based biodiesel a sustainable option for Thailand? J Ind Ecol. 2012;16:541–51.

Singh A, Nigam PS, Murphy JD. Renewable fuels from algae: an answer to debatable land based fuels. Bioresour Technol. 2011;102:10–6.

Slegers PM, Wijffels RH, Van Straten G, Van Boxtel AJ. Design scenarios for flat panel photobioreactors. Appl Energ. 2011;88:3342–53.

Sontheimer A. Alternativen lassen hoffen. J Biogas. 2007;3:42–5.
Stillwell AS, Hoppock DC, Webber ME. Energy recovery from wastewater treatment plants in the United States: a case study of the energy-water nexus. Sustainability. 2010;2:945–62.
Stone KC, Hunt PG, Cantrell KB, Ro KS. The potential impacts of biomass feedstock production on water resource availability. Biores Technol. 2010;101:2014–25.
Stromberg P, Gasparatos A. Biofuels at the confluence of energy security, rural development, and food security: a developing country perspective. In: Gasparatos A, Stromberg P, editors. Socioeconomic and environmental impacts of biofuels. N. Y.: Cambridge Univ. Press; 2012. PP. 3–26.
Suksri P, Moriizumi Y, Hondo H, Yoko W. An introduction of bio-ethanol to thai economy (II)–a survey on sugarcane and cassava processing factories. PhD diss., School of Business and Commerce, Keio University, 2007; Silalertruksa T, Gheewala SH. Environmental sustainability assessment of bio-ethanol production in Thailand. Energy. 2009;34(11):1933–46.
Trostle R. Global agricultural supply and demand: factors contributing to the recent increase in food commodity prices. Washington, DC. US Dept. of Agriculture, Economic Research Service, publication WRS-0801, 30pp. 2008.
Tu Q, Lu M, Yang YJ, Scott D. Water consumption estimates of the biodiesel process in the US. Clean Technol Environ Policy. 2016;18:507–16.
Ugwu CU, Aoyagi H, Uchiyama H. Photobioreactors for mass cultivation of algae. Bioresour Technol. 2008;99:4021–8.
UNEP. Towards sustainable production and use of resources: assessing biofuels. Paris, France: United Nations Environment Programme; 2009.
US-DOE. Theoretical ethanol yield calculator. 2008. http://www1.eere.energygov/biomass/ethanol_yield_calculator.html.
Vasudevan V, Stratton RW, Pearlson MN, Jersey GR, Beyene AG, Weissman JC, et al. Environmental performance of algal biofuel technology options. Environ Sci Technol. 2012;46:2451–9.
Wang W. Cassava production for industrial utilization in China—present and future perspectives [Online]. 2002. http://www.ciat.cgiar.org/asia_cassava/pdf/proceedings_workshop_02/33.pdf.
Wilhelm WW, Johnson JMF, Hatfield JL, Voorhees WB, Linden DR. Crop and soil productivity response to corn residue removal: a literature review. Agron J. 2004;96:1–17.
Xu L, Brilman DW, Withag JA, Brem G, Kersten S. Assessment of a dry and a wet route for the production of biofuels from microalgae: energy balance analysis. Bioresour Technol. 2011;102:113–22.
Yue D, You F, Snyder SW. Biomass-to-bioenergy and biofuel supply chain optimization: overview, key issues and challenges. Comput Chem Eng. 2014;66:36–56.
Ziska LH, Runion GB, Tomecek M, Prior SA, Torbet HA, Sicher R. An evaluation of cassava, sweet potato and field corn as potential carbohydrate sources for bioethanol production in Alabama and Maryland. Biomass Bioenerg. 2009;33:1503–8.

Chapter 4
Impact of Biofuel's Production on Ground Water

Ijaz Bano and Muhammad Arshad

Abstract Groundwater is strategically significant due to its exceeding demand in agriculture, domestic and industrial uses. Global estimates show that approximately 4430 km^3 of fresh water resources are abstracted annually for human consumption. Ground water may contain some unwanted matter with the microbes in its natural form but most of the impurities are being added through human activities. Problem of water pollution is more pronounced in localities where biofuel are produced. The generation of obscene wastewater in bulk is a great environmental apprehension. The chapter will discuss how wastewater generated from biofuel production is deteriorating the ground water quality. The treatment ways of the wastewater will also be discussed. Presence of organic and inorganic compounds in wastewaters release from biofuel production facilities making the ground water unfit for human consumption will be explore.

Keywords Climate change · Ground water · Biofuel · Fresh water quality

4.1 Introduction

Water is the Devine gift, it's ample accessibility in terms of both quantity and quality is vital requirement of each living (Manjunatha et al. 2006). Water for human's consumptions comes across from several sources. Aquatic sources such as river, lakes and water present below the subsoil majorly fulfill the drinking water requirement. In rural areas, most of the water for drinking purpose is sucked through bore wells or hand pumps from underground reservoir (Ray et al. 2000; Garg et al. 2004; Gupta et al. 2014) as ground water is global source of high quality fresh water. Groundwater is strategically significant due to its exceeding demand in

I. Bano · M. Arshad (✉)
Jhang-Campus, University of Veterinary and Animal Sciences, Lahore, Pakistan
e-mail: muhammad.arshad@uvas.edu.pk

I. Bano
e-mail: ijaz.bano@uvas.edu.pk

M. Arshad (ed.), *Perspectives on Water Usage for Biofuels Production*,
https://doi.org/10.1007/978-3-319-66408-8_4

agriculture, domestic and industrial uses. Global estimates show that approximately 4430km^3 of fresh water resources are abstracted annually (Kinzelbach et al. 2003) for human consumption.

The dependence on groundwater is increasing (Ullah et al. 2009). Almost one third of all fresh water extractions, providing approximately 22, 36 and 42% of the water used for industrial, domestic and agricultural purposes comes from the underground reservoirs. Ground water may contain some unwanted matter with the microbes in its natural form but most of the impurities are being added through human activities (Haydar and Qasim 2013). Water pollution is swelling day by day with the unrestrained growth of population. It is considered a cheap source, so least pain is being taken on to conserve its quantity and quality. Mixing of untreated domestic and industrial effluents into water bodies is a common practice now. Adulteration of drinking water quality has become a global issue (Rapant and Krcmova 2007). Problem of water pollution is more pronounced in the developing countries like Pakistan (Haydar and Qasim 2013; Arshad and Ahmed 2016). Just the once adulteration in groundwater occurs; it may become unfeasible for many years. Water quality is uncompromised for mankind (Arain et al. 2008; Dixit and Tiwari 2008) because human health directly depends on it.

Worldwide, above two billion persons rely upon groundwater as the principal source of their drinking water (Li and Merchant 2013; Sampat 2000). Most groundwater depended areas are Northern China, Eastern Europe, Northern India and the US Great Plains. It has been forecasted recently that the collective impacts of increase in population, climatic changes and transformation in land uses will force to higher dependence on groundwater for public water supply (Rosenzweig et al. 2007; Hall et al. 2008).

Climate variability and change impacts the underground reservoirs of water by reduction in recharge replenishment and indirectly through ups and downs in usage. Pace of such impacts is changed by human activity such as land-use change. In recent times, water usage in growing and processing of biomass feedstock to fuels has arisen as most important debate. Production of biofuel may impact both the quantity and quality of water resources that are already under stress. The economic value of the aquifers, especially to agriculture, is highly significant. Continues overdrafts of the aquifers have caused a long-term drop in water levels; some areas have now reached effective depletion (Clark and Peterson 2008).

Production, utilization and management of biofuel interact with the groundwater in amazingly diverse ways. More specific say from hydrogeological science is required to know the links, risk assessment and effective management of such interactions. Sustainable exploitation of biofuel feedstock and their impact on groundwater is needed. The energy utilized for pumping of groundwater and in its use must be estimated (Foster et al. 2015). Biofuel production and combustion processes influence groundwater patterns in a complex way, with a number of direct and indirect effects.

The impacts on groundwater are varied depending on the type of biofuel produced and amount of water input. Changes will come in the quantitative and qualitative stresses on the ground water. Groundwater is important for both economies and ecosystems. In future, groundwater should be managed in a multidisciplinary way in order to provide efficient solutions. The present chapter will seeks to estimate the impact of the emerging biofuel sector on groundwater consumption.

4.2 Repercussion of Groundwater Use

The increased use of ground water for production of biofuel is causing to lower hydraulic heads. It is making additionally expensive to pump water for public use. Schilling et al. (2017) studied utilization of old water for the post-harvest production of ethanol at the plant sites. Chiu et al. (2009) had addressed the issue of groundwater utilization to raise the corn feedstock required for biofuel production. In 2008, 4.5 trillion liters was driven from groundwater to watered corn. Water quality has also been damaged in the aquifer (Vinson et al. 2012). The enhanced water pumping is making the groundwater more vulnerable to deleterious water quality changes for all users.

Further expansion of biofuel production should be built on availability of reliable groundwater sources. Schilling et al. (2017) employed isotopic age dating method to assess the age of groundwater in the aquifer and evaluation of groundwater used. They sampled eight municipal wells for major and minor ions and stable isotopes. According to them groundwater is older than 10,000 years. As the water driven from the C–O aquifer or all uses, including ethanol production, is extracting old groundwater. Upcoming ethanol production projects must be based on availability of sustainable groundwater resources.

4.2.1 Biodiesel Production

Widespread farming of oil crops intended for production of biodiesel broadly use fertilizers and pesticides. Significant irrigation and groundwater is also required in huge quantity. Farming of soya beans in the US with the irrigation of groundwater has increased in last two decades. The impact of such cultivations with the use of groundwater must be evaluated. Liquid biofuel are mixed with gasoline or diesel that keeps toxic compounds like benzene, toluene, ethylbenzene, and xylenes (BTEX) and poly aromatic hydrocarbons (PAHs). Biofuel blended fuel discharges may not degrade through similar fashion in the groundwater due to their certain properties and viscosity.

The aromatic compounds BTEX and PAHs are dangerous to human health in case of any leaking to underground water. The benzene is highly soluble in water; can cause cancer and remains persistence even in anaerobic situation that's present in fuel affected aquifers (Alexander 1999; Alvarez and Illman 2006). Hydrocarbon-contaminated ground waters are normally made safe through biodegradation and natural reduction. Biodiesel keeps various compositions, as it is contrived from different feedstock through trans-esterification process, brings about saturated or unsaturated C_{16}–C_{20} fatty acid methyl esters (FAMEs) (Arshad et al. 2014a). As an example, major portion of rapeseed biodiesel is the methyl ester of erucic acid (Zhang et al. 1998). In soybean biodiesel erucic acid is totally absent and methyl ester of linoleic acid is present upto 55% while castor oil-based biodiesel keeps about 80% methyl ricinoleate. The bacterial community present in groundwater involved in degradation is presented in Table 4.1.

Corseuil et al. (2011) have developed microcosms to study the degradation of biodiesel in groundwater through an experiment. They established that composition of biodiesel significantly affects the biodegradability. Moreover the presence of biodiesel hinders the degradation of benzene and toluene. The decomposition process of ethanol and biodiesel are alike but the variations in the viscosity, solubility, and resulting mobility make the difference. Biotransformation of 80% soybean biodiesel occurred 41 days while only 40% of castor oil biodiesel decomposed in 90 days. High viscosity and less bioavailability reduce the decomposition of castor oil biodiesel. Table 4.2 shows major biochemical reactions involved in decomposition of biodiesel, BTEX and poly aromatic hydrocarbons.

Table 4.1 Groundwater microbes involved in biodegradation

Bacteria	Substrate	References
Janthinobacterium	Poly aromatic hydrocarbons	Lima et al. (2012)
Ralstonia	Fatty acids, alcohols, aromatic compounds	Cramm (2009)
Burkholderia	Methyl esters, acetate, alkanes, chlorinated compounds	Sullivan and Mahenthiralingam (2005), Belova et al. (2006)
Geobacter	Short chain fatty acids, alcohols and mono aromatic compounds	Cord-Ruwisch et al. (1998)
Geobacillus	Organic acids, alcohols, aromatic compounds and carbohydrates	Nazina et al. (2001)
Desulfitobacterium	Halogenated organic compounds, aromatic compounds, organic acids	Villemur et al. (2006), Kunapuli et al. (2010)
Nitrospira	Nitrogenous compounds	Blackbourne et al. (2007)
Pseudomonas	Aromatic compounds	Lima et al. (2012)
Desulfovibrio	Long chain fatty acids, aromatic compounds, organic acids	Allen et al. (2008)
Desulfosporosinus	Sugar, alcohols, monoaromatic compounds,	Ramamoorthy et al. (2006)
Clostridium lll	Long chian fatty acids, organic acids	Hatamono et al. (2007)
Geothrix	Organic acids	Nevin and Lovley (2002)
Anaeromyxobacter	Chlorinated organic compounds, organic acids	He and Sanford (2004)
Pelotomaculm	Organic acids, alcohols	Imachi et al. (2002)
Holophaga	Short chain fatty acids, aromatic compounds	Chauhan and Ogram (2006)

Table 4.2 Major biochemical reactions involved in decomposition of biodiesel, BTEX and poly aromatic hydrocarbons (Ramos et al. 2014)

Name of the compound	The reactions involved
Linoleic acid (C18:2)	$C_{18}H_{31}O_2^- + H^+ + 16H_2O \rightarrow 9CH_3COO^- + 9H^+ 14H_2$ $CH_3COO^- + 9H^+ + 18H_2O \rightarrow 36H_2 + 18CO_2$ $50H_2 + 12.5CO_2 \rightarrow 12.5CH_4 + 25H_2O$ Sum $C_{18}H_{31}O_2^- + 9H_2O \rightarrow 12.5CH_4 + 5.5CO_2$
Benzene	$C_6H_6 + 6H_2O \rightarrow 3CH_3COO^- + 3H^+ + 3H_2$ $3CH_3COO^- + 5H^+ + 10H_2O \rightarrow 12H_2 + 6CO_2$ $15H_2 + 3.75CO_2 \rightarrow 3.75CH_4 + 7.5H_2O$ Sum: $C_6H_6 + 4.5H_2O \rightarrow 3.75CH_4 + 2.25CO_2$
Naphthalene	$C_{10}H_8 + 10H_2O \rightarrow 5CH_3COO^- + 5H^+ + 4H_2$ $5CH_3COO^- + 5H^+ + 10H_2O \rightarrow 10CO_2 + 20H_2$ $24H_2 + 6CO_2 \rightarrow 6CH_4 + 12H_2O$ Sum: $C_{10}H_8 + 8H_2O \rightarrow 6CH_4 + 4CO_2$

The biodegrading ways of BTEX and PAHs in ground water through controlled discharge of biodiesel was observed up to six years. In the first two years, biodiesel became decomposed into acetate and methane. Quantity of benzene remained was observed above the limits till the experiment ended. When it was compared to ethanol blended gasoline, biodiesel/diesel blend release resulted in a shorter BTEX plume, but with higher residual dissolved hydrocarbon concentrations near the source zone (Ramos et al. 2016).

Fedrizzi et al. (2015) stated that increasing demand of biodiesel will lead to the chances of contamination in aquifer and current microbial methods may be successful to treat through indigenous microbial community. They conducted an experiment by releasing 100 L of palm biodiesel to check the probability of bioremediation. Contamination following the peroxidation simulated the microbial community and responded positively to contamination. Decomposition of palm biodiesel occurred as bioremediation technique worked well.

4.2.2 Biogas Production

Biofuel production is promoted by globally to fight climatic change and to ensure better future energy supplies. Biogas produced through anaerobic digestion of biomass and manure is an example of renewable energy source, which may reduce global warming impacts (Piątek et al. 2016). The efficient use of biogas residues has become a challenge. It is produced in bulk and must be recycled. The field applications of biogas residues keep higher N-leaching risk to the ground water. In case of crop sowing as maize, the organic fertilizer is commonly applied (Zhao et al. 2006) that leaches to the ground water. Biodigestate has a similar leaching risk to that of animal slurry. The risk of nitrate leaching was substantially higher for maize than for grassland (Albert et al. 2016).

4.2.3 Bioethanol Production

The impact of the grain-derived ethanol industry on cropping systems, and the associated water and nutrient demand and the runoff of chemicals into surface and underground water bodies are important, but very complex issues (Delucchi 2010; Babel et al. 2011). Manufacturing of ethanol from sugarcane has become an imperative unit of sugar industry worldwide (Arshad et al. 2008, 2011, 2014b, 2017; Arshad 2010). But, the generation of obscene wastewater in bulk known as stillage is a great environmental apprehension. For every liter of alcohol produced, 12–15 L of distillery effluent are generated. The treatment of this wastewater before disposal is regulatory; however, owing to the high effluent treatment costs and due to elaborate physico-chemical methods, partial treatment is carried out and huge quantities of the effluent is either stored in lagoons, unlined tanks or let out into the surface water bodies or streams, thereby adversely affecting the good quality water resources and environment. The present effort focuses on the effects of distillery effluents on the groundwater quality in vicinity of distillery.

All the industries discharge their waste water contaminants in waste water are usually a complex mixture of organic and inorganic compounds which make the natural water unfit for human consumption. Not only this, hand pump water of nearby colonies also gets polluted due to the seepage of contaminated water. The discharge of untreated liquid effluents above the National Environmental Quality Standards (NEQS) for industrial effluents may affect quality of soil, groundwater and receiving water bodies (Baskran et al. 2009; Rehman et al. 2009). The toxic materials may enter the food chain and cause problem as the effluents of the sugar industry enter the water bodies and effect human health. The effects of have been reported (Ayyasamy et al. 2008; Khan et al. 2013).

The biofuel blended fuel discharge that can affect groundwater may occur and such events especially regarding bioethanol are increasing. Therefore, it is necessary to understand the process of biodegradation and impacts on quality of groundwater. Response of microbes towards such releases must be explored. As and when any ethanol amended fuel discharge occurs, it penetrates as a non-aqueous phase liquid (NAPL) through the unsaturated zone to the water table and forms a floating NAPL pool at the water table (Freitas and Barker 2011). Ethanol likes to mount at the water table interface (Molson et al. 2008; Capiro et al. 2007). In case of high ratio ethanol amended fuels as 95% ethanol and 5% gasoline, the released fuel will possibly travel on, as water miscible phase. It will split into two phases, as the fuel becomes diluted, precipitating a new NAPL phase along its path (Stafford et al. 2009; Yu et al. 2009). Water present in pores keeps high ethanol ratio with presence of hydrocarbons due to their enhanced solubility in the presence of ethanol (Chen et al. 2008; He et al. 2011).

Various areas having microbial activity will be likely to develop.

1. An area where anaerobic activity will occur in contaminated plume in the saturated zone. An aerobic degradation can occur at the peripheries of the plume.

2. The region of high anaerobic activity in the capillary zone.
3. The area in the unsaturated zone where aerobic degradation of methane is predominant.

Ethanol asserts minimal adverse effects on human health in case of direct exposure, but it can enhance impact of other toxic fuel components such as benzene, toluene, ethylbenzene and xylenes (BTEX) through hindrance in their bio-decomposition process (Powers et al. 2001). The mechanisms by which ethanol impacts BTEX degradation has been presented in Table 4.3. For bio-decomposition process of ethanol and BTEX, sequences of bio-conversions are performed by a variety of microbes working in presences or absence of oxygen (Fuchs et al. 2011). A high BOD is required in areas where high concentration of ethanol is present especially in newly affected groundwater. Yet, aerobic microbial activity is also significant as oxygenases convert BTEX to catechol type compounds or byproducts such as acetyl-CoA, acetaldehyde and pyruvic acid (Alvarez and Illman 2005). Aerobic microbes can metabolize ethanol into the acetyl-CoA via acetaldehyde and acetate, while BTEX are firstly converted to a common aromatic intermediate, benzyl-CoA (Fuchs et al. 2011). In anaerobic digestions acetate are produced that's finally changed into CH_4 and CO_2.

Higher amounts of ethanol are very toxic to microbes and phospholipids can be dissolves and causes the disintegration of cell membrane (Ingram and Buttke 1984). With the loss of cell membrane, ethanol may enter into the cell and can denature the enzymes. Higher quantities of ethanol could stop the biosynthesis of DNA, RNA and proteins.

Decomposition of biothanol may result in comparatively higher methane ratio in groundwater (Freitas et al. 2010; Jewell and Wilson 2011). It can cause an explosion risk due to high accumulation (Bjerketvedt et al. 1997). Such explosion accidents have been reported at landfill sites (Williams and Aitkenhead 1991;

Table 4.3 The mechanisms by which ethanol impacts microbial decomposition of BTEX (Ma et al. 2011)

Mechanisms	System affected	Effects on BTEX degradation
Catabolite repression	Gene expression	–
Metabolic flux dilution	Metabolism	–
pH decrease	Cell physiology	–
Ethanol toxicity	Physiology and metabolism	–
Fortuitous growth of BTEX degraders	Community structure	+
Genotypic dilution	Community structure	–
Growth of syntrophic microorganisms	Community structure	+
Increase richness and diversity	Community structure	+
Electron acceptor/nutrients depletion	Metabolism, kinetics	–
Thermodynamic inhibition due to VFAs accumulation	Metabolism, kinetics	–

Kjeldsen 1996). Volatile fatty acids (VFAs) produced from ethanol biodegradation can generate odor that may confront the aesthetic quality of aquifer. Odor is high meaningful for the public's opinion regarding quality of drinking water (EPA 2002). In an experimental study, the odor level in the VFAs affected aquifer was 350 times higher; the major odor contributor was butyric acid (Ma et al. 2011).

Spalding et al. (2011) examined E95 samples collected from a spill for BTEX, ethanol, methane and acetate. Bioethanol attentiveness was only limited to the released area. An anaerobic zone was also present in aquifer. BTEX appeared to be persistent. Methane appeared to be generated within the capillary fringe and very shallow groundwater and migrate laterally. Methane's high oxygen demand promotes anaerobic conditions within the shallow groundwater. Estimated and measured methane soil gas concentrations exceeded the lower explosive limit.

Regarding effects of ethanol amended biofuel following precautions may be considered.

1. Concentrations of acetates must be observed in aquifer as it interfere the thermodynamic possibility of anaerobic BTEX disintegration and may suppress inducible enzymes.
2. Bio-decomposition of ethanol can produce CH_4 that may cause an explosion risk. Long-term observation of CH_4 in groundwater and soil gas near the source zone should be considered.
3. Bioethanol amended fuels may result in less aerobic offsetting of BTEX vapors. It may risk the potential of BTEX vapor interruption into overlying buildings. The vapor interruption risk must be considered.
4. Normally bioethanol mount up and travels parallel within the capillary fringe, monitoring of ethanol should focus on this zone.
5. Bioethanol may hinder BTEX decomposition. It must be well-thought-out for source removal.

4.3 Biofuel Production and Irrigation

Commonly used commercial biofuel are mostly produced from sugars keeping crops like sugarcane, sugar beet, corn and oil bearing crops like canola, soybean oil palm. The biofuel crops devoted area is comparatively very small and irrigation water drafting to grow such crops is also very small globally (Tyagi 2015). Water shortage problems are arising in regions. Quantity of irrigation water extracted from groundwater in the year 2000 has been shown in Table 4.4.

The appropriate cropland is randomly spread worldwide. Two biofuel producing leaders are US and Brazil and China, India, and the EU will also be included in this list (De Fraiture et al. 2008). In 2030, above 3% cropland and irrigation water drafting over 4% are forecasted for biofuel production (Eisentraut 2013). There must be studied about water availability, water footprint, and virtual water quality. As the biofuel production carries on inflating, the requirement of energy crops will

Table 4.4 Extraction of ground water (km^3/year) for irrigation purpose in the year 2000 (Renault and Wallender 2000; De Marsily and Abarea-del-Rio 2016)

Country	Extraction of groundwater in a year	Over drafting as compared to natural recharge	Annual use of irrigation
India	190	71	600
USA	115	32	204
China	97	22	403
Pakistan	55	37	183
Iran	53	27	59
Mexico	38	11	71
Total	548	200	1520
Planet Earth	734	256	2510

be more amplified. Such demand has shifted worldwide crop producers to switch production from many traditional crops, such as wheat and grain sorghum, to high energy crops, most notably water- and input-intensive corn, which are more suitable for biofuel production. While this substitution has created volatility in commodity markets, it has also created an additional strain on water resources in many irrigated agricultural regions.

4.3.1 Decline in Groundwater Due to Water Over Drafting for Corn and Soybeans

Commercial value of corn and soybean has replaced cotton land with increase withdrawal of water. The increase in corn prices driven by demand for ethanol-based biofuel resulted in a 47% reduction in cotton acreage concurrent with a 288% increase in corn acreage in 2007 relative to 2006.

Withdrawal of groundwater to grow a crop varies for each and every season depends upon on weather conditions, irrigation methods. As an example, the pumping of water for corn amplified up to 566 million m^3 in 2007, while water suction for cotton crop decreased to 282 million m^3 during the same timeframe. Due to timely rains, there was 30% less water consumption in 2007 as compared to 2006 (Welch et al. 2010). Table 4.5 shows the ground water depletion in Km^3/Yr.

Table 4.5 Approximate ground water depletion (km^3yr^{-1}) in various continents and worldwide (Taylor et al. 2013)

Region	Flux-based method	Volume-based method
World	204 ± 30	145 ± 39
Asia	150 ± 25	111 ± 30
Africa	5.0 ± 1.5	5.5 ± 1.5
North America	40 ± 10	26 ± 7
South America	1.5 ± 0.5	0.9 ± 0.5
Australia	0.5 ± 0.2	0.4 ± 0.2
Europe	7 ± 2	1.3 ± 0.7

4.4 Impacts on Groundwater Quality

Although improved on-farm irrigation efficiency and water recycling within biorefineries may not increase the amount of water available downstream, they may improve downstream water quality by reducing the throughput of water into production. By reducing required withdrawals and increasing consumptive use, each measure leaves more water at the source and reduces discharges of unconsumed water back to the source (Huffaker 2010). The alcohol distilleries associated with sugar industry, produce 15–16 L of spent wash for every liter of alcohol. The spent wash is highly loaded with BOD and COD and could create the environmental problems and danger to human health, if it is allowed to interact with soil or water bodies (Kumar and Gopal 2001; Workocha 2011).

The spent wash may percolate the sub soil to affect the ground water. The quality of spent wash located in unlined evaporation ponds and their effect on the groundwater used for drinking purposes and irrigation around the evaporation ponds must be examined. The quality of the groundwater is compared with reference samples of ground water collected at a distance of about 4–5 km from the1evaporation ponds. Preliminary results of impact of evaporation ponds of ethanol distillery spent wash on groundwater were presented (Khuhawar et al. 2011; Mahar et al. 2012, 2013).

4.4.1 Impact of Increased Agrichemical Application

Wilhelm et al. (2010) has quoted a mathematical model regarding water-quality to explore the increase in fertilizer usage rates as consequence of increased corn production. Their model was calibrated according to existing scenario to evaluate fluxes of water and chemicals from agricultural fields to groundwater (Green et al. 2009). There was nitrate/nitrogen contamination in shallow groundwater due to leaching of chemical nitrogen.

4.4.2 *Influence of Climatic Uncertainty on Groundwater Remediation and Restoration*

The manifestation of shifting climatic conditions, both short term and long term, will have a strong influence on the design of reliable and protective groundwater contamination mitigation and water resource protection measures. Forecasts of our hydraulic future not as easily rely on historical patterns. In many parts of North America, Australia, South America, and Europe precipitation and runoff patterns are changing and extreme climatic events (droughts, storms, floods) are expected to become more frequent.

Rising temperatures and sea levels are being observed and are forecast to become more prevalent. These anticipated conditions relate strongly to water resource availability, distribution, and vulnerability, and they challenge our ability to develop effective and robust contaminant clean up and long-term and reliable protection measures. Likely perturbations to groundwater level and flow conditions must be assessed; groundwater remediation measures must account for the uncertainty in near shore and continental recharge conditions; and, the fate of contaminant plumes will arguably become less predictable as climatic variability affects both hydraulic and chemical conditions in the subsurface (Anderson et al. 2015). The convention of assessing and designing "30-year" groundwater protection and clean up remedies no longer seems relevant; the reality is that our remedial strategies could benefit from adaptive approaches to both passive and active restoration concepts. This will require a better understanding of how our remedial systems age, either acutely or chronically, with changing hydraulic conditions, and how we can engineer and monitor remedial systems using reliability-based methods under different hydraulic scenarios.

While climatic and hydrologic changes demand consideration of technical innovations to promote reliable and protective water resource clean up measures, many key stressors will impact the economic evaluation of which measure makes the most sense for a given situation (Chen et al. 2015b). Theses stressors will include toxicological changes for given chemicals as geochemical conditions in a groundwater system as hydraulic shift; and may include usability of groundwater depending on the availability of surface water in drought conditions or in periods of rapid urbanization. Environmental stresses can increase due to climatic stress, and the resulting impact on consumer goods, agriculture, and ecosystem needs will provide even greater stress on the distribution of water resources, which will impact the magnitude of the economic investment into a groundwater cleanup method. The economic consideration is important to assuring that that the groundwater resource remains a viable asset as determined by both environmental and economic net benefit analysis for the myriad of environmental and human uses of a given water resource (Jones 2014).

4.4.3 Impact of Crop Residue Use in Biofuel Production

Harvesting corn residues (stover) in water-constrained areas of the United States (US) Corn Belt may trigger losses in stored soil water through increased evaporation (short term effect), and reduced soil organic carbon (SOC) which lessens the water holding capacity of the soil (long term effect) (Sesmero 2014). This study develops a model of stover supply and derived demand for irrigation water accounting for the adverse effects of stover removal on soil water. A cost-minimizing plant takes into account this supply response when deciding the price it will offer for stover. Stover price, total groundwater consumption and SOC reduction associated with different levels of biofuel production are calculated and compared to a baseline without biofuel so that implications of biofuel development on these natural resources are quantified. We also quantify the effect of water conservation policies on the cost of producing biofuel.

4.4.4 Leaching of Nitrogenous Compounds into Ground Water

Nitrogen (N) is vital element for life on earth (Erisman et al. 2015). When excess N leached into groundwater then exposure to this excess N can result in ecological and human health impacts such as fish kills, human disease and birth defects (Johnson et al. 2010; Ward et al. 2005). As agricultural practices are major source of N released into the environment and growing the corn needs huge quantity of N fertilization (Ribaudo et al. 2011; Sobota et al. 2013). The increasing need of corn for biofuel production has consequently increased requirement of N fertilizer that may sustenance the risk to human health due to drinking of contaminated ground water (Brender et al. 2013).

Garcia et al. 2017 studied a coupled modeling system to simulate the effects of different corn production situations. N-loadings to groundwater were used in a statistical model to describe as a function of a variety of environmental variables. Differences among the scenarios and among the high groundwater nitrate-N concentrations ($\geq$ 5 mg/L) were used. The difference in the averaged N-fertilizer rates for each grid cell, as well as the difference in the fertilizer rate for just the increased corn cropland within each grid cell. Wethen used these "biofuel-only" fertilizer rates to predict groundwater nitrate-N. little difference in the overall averaged data between the scenarios for Ksat, AFOs or unconsolidated aquifers. There is evidence, however, of corn production expansion in response to increased biofuel demand, with a 27% increase in the number of grid cells projected to contain irrigated grain-corn croplands between the 2002 and 2022 CORN domains.

4.4.5 Soil Contamination

The future fuels (biofuel) must be evaluated in respect of their potential to harm environment especially to water resources. Possibility of chemicals leakage from biofuel affected soils must be explored to evaluate the hazard to groundwater (Efroymson and Dale 2015). In this way the sorption behavior of biofuel components must be understood. Chen et al. (2015a) formulated batch and column experiments to assess the sorption and desorption performance of fuel constituents. It was observed that when ethanol contents in ethanol-blended gasoline exceeded 25%, enhanced desorption of the aromatic constituents to water was observed. In case of biodiesel, the sorption of such materials remained unaffected. Organic carbon contents of the soil also affected desorption of target compounds. Their findings can provide the basics to predict the fate and transport of hydrophobic organic compounds and sorption behavior of biofuel components for future assessments of the impacts of biofuel.

4.5 Alternative Water Resources

Search for substitute water resources is on rise due to apprehensions of high water demand for crop/ algal biofuel (Beal et al. 2012). The production of biofuel must be cost effective and not compatible with food. Table 4.6 presents the water requirements for production of food (m^3t^{-1}). The viability of consuming effluent of municipal wastewater treatment plant is being explored. Use of saline water, seawater, and dairy wastewater as water resources is being examined. Such resources can be good source of nutrient (Chiu et al. 2013; Venteris et al. 2013; Kothari et al. 2012; Anderson et al. 2012; Roberts et al. 2013) and able to sustain the algal growth with the provision of sufficient nutrients (Kothari et al. 2012; Anderson et al. 2012). Linkage of biofuel production with municipal wastewater treatment can make easier use of nutrient with the water reuse and will definitely save energy.

Table 4.6 Average water requirement (m^3 t^{-1}) for production of various food items (De Marsily and Abarea-del-Rio 2016)

Vegetal product	Water needed (m^3 t^{-1})	Animal product	Water need (m^3 t^{-1})
Vegetable oil	5000	Beef	13,000
Rice	1500–2000	Poultry	4100
Wheat	1000	Eggs	2700
Corn	700	Milk	800
Citrus fruits	400		
Vegetables	200–400		
Potatoes	100		

There may be some mismatch among water requirements for algal growth and the availability of municipal waste water (MWW) treatment plant effluent. With the use spatially and temporally available MWW as a sole source of water, 8.6 billion liters of bio-oil can be produced annually with a freshwater blue water footprint that is almost nil (Chiu et al. 2013). Techno-economic analysis further indicated that when brackish and saline water are adopted as backup water for open pond operation, the consumption of freshwater for the production of algae fuel can be comparable to that of petroleum-derived fuels (Vasudevan et al. 2012).

4.6 Conclusion

The resilience of home, farm and industrial application of fresh water in the face of climate changes can be enhanced through safeguarding the ground water. As present upsurge of biofuel production will continue in future, impact of climatic changes on ground water may be greatest through indirect effects on water needed for irrigation. There is much uncertainty regarding impacts of climatic variations (Bates et al. 2008), but fluctuations of precipitation pattern with temperature extremes are much believable (Allan and Soden 2008). Intense rainfall events may follow the drought periods. At first, such fluctuations in climatic conditions may have impact on ground water by increase in water needed for irrigation due to decreased recharge and increased discharge. Two thirds irrigated area in 1995 will seek more water for irrigation in 2070 (Döll 2002) due to variations in climatic conditions. Broad water management approaches to integrate the ground water and surface water sources, can minimize human vulnerability to climate extreme variations.

References

Albert C, Hermes J, Neuendorf F, von Haaren C, Rode M. Assessing and governing ecosystem services trade-offs in agrarian landscapes: the case of biogas. Land. 2016;5:1.

Alexander M. Biodegradation and bioremediation. 2nd ed. San Diego, California: Academic Press; 1999.

Allan RP, Soden BJ. Atmospheric warming and the amplification of precipitation extremes. Science. 2008;321:1481–4.

Allen TD, Kraus PF, Lawson PA, Drake GR, Balkwill DL, Tanner RS. *Desulfovibrio carbinoliphilus* sp., a benzyl alcohol-oxidizing, sulfate-reducing bacterium isolated from a gas condensate-contaminated aquifer. Int J Syst Evol Microbiol. 2008;58:1313–7.

Alvarez PJ, Illman WA. Bioremediation and natural attenuation: process fundamentals and mathematical models. Hoboken, New Jersey, USA: Wiley-Interscience; 2005. p. 28.

Alvarez PJ, Illman WA, eds. Bioremediation and natural attenuation: Process fundamentals and mathematical models. In: Schnoor JL, Zehnder A, editors. Environmental science and technology. Hoboken, New Jersey: Wiley; 2006.

Anderson MP, Woessner WW, Hunt RJ. Applied groundwater modeling: simulation of flow and advective transport. Academic Press; 2015 Aug 13.

Andersson V, Broberg S, Hackl R. Integrated algae cultivation for municipal wastewater treatment and biofuels production in industrial clusters. In: Proceedings of WREF, Denver, USA. 2012 May:13–7.

Arain MB, Kazi TG, Jamali MK, Jalbani N, Afridi HI, Shah A. Total dissolved and bioavailable elements in water and sediment samples and their accumulation in *Oreochromis mossambicus* of polluted Manchar Lake. Chemosphere. 2008;70:1845–56.

Arshad M. Bioethanol: A sustainable and environment friendly solution for Pakistan. A Scientific J. COMSATS–Sci. Vision. 2010;16–7.

Arshad M, Ahmed S. Cogeneration through bagasse: a renewable strategy to meet the future energy needs. Renew Sust Energ Rev. 2016;54:732–7.

Arshad M, Khan ZM, Shah FA, Rajoka MI. Optimization of process variables for minimization of byproduct formation during fermentation of blackstrap molasses to ethanol at industrial scale. Lett Appl Microbiol. 2008;47:410–4.

Arshad M, Zia MA, Asghar M, Bhatti H. Improving bio-ethanol yield: Using virginiamycin and sodium flouride at a Pakistani distillery. Afr J Biotechnol. 2011;10:11071.

Arshad M, Adil M, Sikandar A, Hussain T. Exploitation of meat industry by-products for biodiesel production: Pakistan's perspective. Pakistan J Life Soc Sci. 2014a;12:120–5.

Arshad M, Ahmed S, Zia MA, Rajoka MI. Kinetics and thermodynamics of ethanol production by *Saccharomyces cerevisiae* MLD10 using molasses. Appl Biochem Biotechnol. 2014b;172:2455–64.

Arshad M, Hussain T, Iqbal M, Abbas M. Enhanced ethanol production at commercial scale from molasses using high gravity technology by mutant *S. cerevisiae*. Brazilian J Microbiol. 2017. http://doi.org/10.1016/j.bjm.2017.02.003.

Ayyasamy PM, Yasodha R, Rajakumar S, Lakshmanaperumalsamy PK, Rahman PK, Lee S. Impact of sugar factory effluent on the growth and biochemical characteristics of terrestrial and aquatic plants. Bull Environ Contam Toxicol. 2008;81:449–54.

Babel MS, Shrestha B, Perret SR. Hydrological impact of biofuel production: a case study of the Khlong Phlo watershed in Thailand. Agric Water Manag. 2011;101:8–26.

Baskaran S, Ransley T, Brodie RS, Baker P. Investigating groundwater–river interactions using environmental tracers. Australian J Earth Sci. 2009;56:13–9.

Bates BC, Kundzewicz ZW, Wu, S, Palutikof JP. Climate change and water technical paper of the intergovernmental panel on climate change VI (IPCC, 2008).

Beal CM, Hebner RE, Webber ME, Ruoff RS, Seibert AF, King CW. Comprehensive evaluation of algal biofuel production: experimental and target results. Energies. 2012;5:1943–81.

Belova SE, Pankratov TA, Dedysh SN. Bacteria of the genus *Burkholderia* as a typical component of the microbial community of *Sphagnum* peat bogs. Microbiol. 2006;75:90–6.

Bjerketvedt D, Bakke JR, van Wingerden K. Gas explosion handbook. J Hazard Mater. 1997;52:1–150.

Blackbourne R, Vadivelu VM, Yuan Z, Keller J. Kinetic characterisation of an enriched *Nitrospira* culture with comparison to *Nitrobacter*. Water Res. 2007;41:3033–42.

Brender JD, Weyer PJ, Romitti PA, Mohanty BP, Shinde MU, Vuong AM, et al. Prenatal nitrate-N intake from drinking water and selected birth defects in offspring of participants in the national birth defects prevention study. Environ Health Perspect. 2013;121:1083–9.

Capiro NL, Stafford BP, Rixey WG, Bedient PB, Alvarez PJJ. Fuelgrade ethanol transport and impacts to groundwater in a pilotscale aquifer tank. Water Res. 2007;41:656–64.

Chauhan A, Ogram A. Fatty acid-oxidizing consortia along a nutrient gradient in the Florida Everglades. Appl Environ Microbiol. 2006;72:2400–6.

Chen CS, Lai YW, Tien CJ. Partitioning of aromatic and oxygenated constituents into water from regular and ethanol blended gasolines. Environ Pollut. 2008;156:988–96.

Chen CS, Shu YY, Wu SH, Tien CJ. Assessing soil and groundwater contamination from biofuel spills. Environ Sci. 2015;17:533–42.

Chen Y, Ale S, Rajan N, Morgan CL, Park J. Hydrological responses of land use change from cotton (*Gossypium hirsutum* L.) to cellulosic bioenergy crops in the Southern High Plains of Texas, USA. Gcb Bioenergy. 2015b.

Chiu Y-W, Walseth B, Suh S. Water embodies in bioethanol in the United States. Environ Sci Technol. 2009;43:2688–92.

Chiu Y-W, Wu M. Considering water availability and wastewater resources in the development of algal bio-oil. BioFPR. 2013;7:406–15.

Clark MK, Peterson JM. Biofuel boom, aquifer loom. In: Selected paper prepared for presentation at the American agricultural economics association annual meeting, Orlando, Florida July. 2008.

Cord-Ruwisch R, Lovley DR, Schink B. Growth of *Geobacter sulfurreducens* with acetate in syntrophic cooperation with hydrogen-oxidizing anaerobic partners. Appl Environ Microbiol. 1998;64:2232–6.

Corseuil HX, Monier AL, Gomes AP, Chiaranda HS, do Rosario M, d Alvarez PJ. Biodegradation of soybean and castor oil biodiesel: implications on the natural attenuation of monoaromatic hydrocarbons in groundwater. Ground Water Monit Remed. 2011;31:111–8.

Cramm R. Genomic view of energy metabolism in *Ralstonia eutropha* H16. J Mol Microbiol Biotechnol. 2009;16:38–52.

De Fraiture C, Giordano M, Liao Y. Biofuels and implications for agricultural water use: blue impacts of green energy. Water Policy. 2008;10:67–81.

De Marsily G, Abarca-del-Rio R. Water and food in the twenty-first century. Surveys Geophy. 2016;37:503–27.

Delucchi MA. Impacts of biofuels on climate change, water use, and land use. Ann N Y Acad Sci. 2010;1195:28–45.

Dixit S, Tiwari S. Impact assessment of heavy metal pollution of Shahpuralake, Bhopal. India. Int J Environ Res. 2008;2:37–42.

Döll P. Impact of climate change and variability on irrigation requirements: a global perspective. Clim Change. 2002;54:269–93.

Efroymson RA, Dale VH. Environmental indicators for sustainable production of algal biofuels. Ecol Ind. 2015;28:49:1–3.

Eisentraut A. The biofuel and bioenergy roadmaps of the international energy agency. In Dallemand JF, Gerbens-Leenes PW, editors. Bioenergy and water, JRC Technical Report, The Netherlands. 2013.

EPA: National Secondary Drinking Water Regulations. Washington, DC, USA: U.S. Environmental Protection Agency. 2002. http://www.gpo.gov/fdsys/pkg/CFR-2002-title40-vol1/pdf/CFR-2002-title40-vol1.pdf.

Erisman, JW, Enrico D, Martin Van D, Nadejda S, and Martijn S. Impacts Ecosystems. (2015).

Fedrizzi F. Degradation of biodiesel in groundwater. http://www.pt.congresso2015.solabiaa.com.br/resources/anais/4/1447041643_ARQUIVO_9999299(okg).pdf.

Foster S, Tyson G, Ferguson G, Younger P, Bath A, Evans R, et al. The energy sector and groundwater. Bruce Misstear and John Chilton. 2015.

Freitas JG, Barker JF. Oxygenated gasoline release in the unsaturated zone – Part 1: source zone behavior. J Contam Hydrol. 2011;126:153–66.

Freitas JG, Fletcher B, Aravena R, Barker JF. Methane production and isotopic fingerprinting in ethanol fuel contaminated sites. Ground Water. 2010;48:844–57.

Fuchs G, Boll M, Heider J. Microbial degradation of aromatic compounds-from one strategy to four. Nat Rev Microbiol. 2011;9:803–16.

Garcia V, Cooter E, Crooks J, Hinckley B, Murphy M, Xing X. Examining the impacts of increased corn production on groundwater quality using a coupled modeling system. Sci Total Environ. 2017;586:16–24.

Garg M, Kavita VK, Malik RA. Groundwater quality in some villages of Haryana India: focus on fluoride and fluorosis. J Hazard Mater. 2004;106B:85–97.

Green CT, Heather W, Richard C. Multi-tracer analysis of vertical nitrate fluxes in the Mississippi River Valley alluvial aquifer [abs.]. In: Eos transactions of the American Geophysical Union. 2009;90:H31C–0799.

Gupta GS, Gupta MK, Alpana G, Anjani G. Assessment of physic-chemical characteristics of hand pumps water of Banda city Indian. J Environ Prot. 2014;34:51–4.

Hall ND, Stuntz BB, Abrams RH. Climate change and freshwater resources. Nat Resour Environ. 2008;22:30–5.

Hatamono M, Imachi H, Yashiro Y, Ohashi A, Harada H. Diversity of anaerobic microorganisms involved in long-chain fatty acid degradation in methanogenic sludges as revealed by RNA-based stable isotope probing. Appl Environ Microbiol. 2007;73:4119–27.

Haydar S, Qasim MM. A study of water quality of Sargodha city. Pak J Engg Appl Sci. 2013;13:110–7.

He Q, Sanford RA. Acetate threshold concentrations suggest varying energy requirements during anaerobic respiration by *Anaeromyxobacter dehalogenans*. Appl Environ Microbiol. 2004;70:6940–3.

He XH, Stafford BP, Rixey WG. Ethanol-enhanced dissolution of a residually trapped synthetic gasoline source. Ground Water Monit Remed. 2011;31:61–8.

Huffaker R. Protecting water resources in biofuels production. Water Policy. 2010;12:129–34.

Imachi H, Sekiguchi Y, Kamagata Y, Hanada S, Ohashi A, Harada H. *Pelotomaculum thermopropionicum* gen. nov., sp. nov., an anaerobic, thermophilic, syntrophic propionate-oxidizing bacterium. Int J Syst Evol Microbiol. 2002;52:1729–35.

Ingram LO, Buttke TM. Effects of alcohols on microorganisms. Adv Microb Physiol. 1984;25:253–300.

Jewell KP, Wilson JT. A new screening method for methane in soil gas using existing groundwater monitoring wells. Ground Water Monit Rremediation. 2011;31:82–94.

Johnson PTJ, Townsend AR, Cleveland CC, Glibert PM, Howarth RW, McKenzie VJ, et al. Linking environmental nutrient enrichment and disease emergence in humans and wildlife. Ecol Appl. 2010;20:16–29.

Jones WR, Spence MJ, Bowman AW, Evers L, Molinari DA. A software tool for the spatiotemporal analysis and reporting of groundwater monitoring data. Environ Model Softw. 2014;55:242–9.

Khan MU, Malik RN, Muhammad S. Human health risk from heavy metal via food crops consumption with wastewater irrigation practices in Pakistan. Chemosphere. 2013;93:2230–8.

Khuhawar MY, Baloch MA, Jahangir TM, Mahar MT, Majidano SA. Impacts of evaporation ponds of ethanol distillery spent wash on underground water. Pak J Chem. 2011;1:10–8.

Kinzelbach W, Bauer P, Siegfried T, Brunner P. Sustainable groundwater management-problems and scientific tools. Institute for Hydromechanics and water resources management, ETH, Zurich, Switzerland. 2003;26:279–83.

Kjeldsen P. Landfill gas migration in soil. In Landfilling of Waste: Biogas. Edited by Christensen TH, Cossu R, Stegmann R. London, UK: E & FN Spon; 1996.

Kothari R, Pathak V, Kumar V, Singh DP. Experimental study for growth potential of unicellular alga Chlorella pyrenoidosa on dairy wastewater: an integrated approach for treatment and biofuel production. Bioresour Technol. 2012;16:466–70.

Kumar S, Gopal K. Impact of distillery effluent on physiological consequences in the freshwater teleost Channa punctatus. Bull Environ Contam Toxicol. 2001;66:617–22.

Kunapuli U, Jahn MK, Lueders T, Geyer R, Hermann JH, Meckenstock RU. *Desulfitobacterium aromaticivorans* sp. nov. and *Gebacter toluenoxydans* sp. nov., iron-reducing bacteria capable of anaerobic degradation of monoaromatic hydrocarbons. Int J Syst Evol Microbiol. 2010;60:686–95.

Li R, Merchant JW. Modeling vulnerability of groundwater to pollution under future scenarios of climate change and biofuels-related land use change: a case study in North Dakota. USA. Sci Total Environ. 2013;447:32–45.

Lima G, Parker B, Meyer J. Dechlorinating microorganisms in a sedimentary rock matrix contaminated with a mixture of VOCs. Environ Sci Technol. 2012;46:5756–63.

Ma J, Xiu Z, Monier A, Mamonkina I, Zhang Y, He Y, et al. Aesthetic groundwater quality Impacts from a continuous pilot-scale release of an ethanol blend. Ground Water Monit Remed. 2011;31:47–54.

Mahar MT, Khuhawar MY, Baloch MA, Jahangir TM. Effects of spent wash of ethanol industry on groundwater: a case study of Rahimyar Khan district,Pakistan. J Environ Sci Water Resour. 2012;1:85–94.

Mahar MT, Khuhawar MY, Jahangir TM, Baloch MA. health risk assessment of heavy metals in groundwater: The effect of evaporation ponds of distillery spent wash. J Environ Sci Eng. 2013;2:166.

Manjunath AV, Chengapp PG, Chandrakanth MG. Externalities due to sand mining and distillery effluent in water streams of india. J Glob Economy. 2006;2.

Molson J, Mocanu M, Barker J. Numerical analysis of buoyancy effects during the dissolution and transport of oxygenated gasoline in groundwater. Water Resour Res. 2008;44:W07418.

Nazina TN, Tourova TP, Poltaraus AB, Novikova EV, Grigoryan AA, Ivanova AE, et al. Taxonomic study of aerobic thermophilic bacilli: descriptions of *Geobacillus subterraneus* gen. nov., sp. nov. and *Geobacillus uzenensissp* nov. from petroleum reservoirs and transfer of *Bacillus stearothermophilus*, *Bacillus thermo-catenulatus*, *Bacillus thermoleovorans*, *Bacillus kaustophilus*, *Bacillus thermoglucosidasius* and *Bacillus thermodenitrificans* to *Geobacillus* as the new combinations *G. stearothermophilus*, *G. thermocatenulatus*, *G. thermoleovorans*, *G. kaustophilus*, *G. thermoglucosidasius* and *G. thermodenitrificans*. Int J Syst Evol Microbiol. 2001;51:433–46.

Nevin KP, Lovley DR. Mechanisms for accessing insoluble Fe(III) oxide during dissimilatory Fe (III) reduction by *Geothrix fermentans*. Appl Environ Microbiol. 2002;68:2294–9.

O'Sullivan LA, Mahenthiralingam E. Biotechnological potential within the genus *Burkholderia*. Lett Appl Microbiol. 2005;41:8–11.

Piątek M, Lisowski A, Kasprzycka A, Lisowska B. The dynamics of an anaerobic digestion of crop substrates with an unfavourable carbon to nitrogen ratio. Bioresour Technol. 2016;216:607–12.

Powers SE, Hunt CS, Heermann SE, Corseuil HX, Rice D, Alvarez PJJ. The transport and fate of ethanol and BTEX in groundwater contaminated by gasohol. Crit Rev Environ Sci Technol. 2001;31:79–123.

Rahman MM, Naidu R, Bhattacharya P. Arsenic contamination in groundwater in the Southeast Asia region. Environ Geochem Health. 2009;31:9–21.

Ramamoorthy S, Sass H, Langner H, Schumann P, Kroppenstedt RM, Spring S, Overmann J, Rosenzweig RF. *Desulfosporosinus lacus* sp. nov., a sulfate-reducing bacterium isolated from pristine freshwater lake sediments. Int J Syst Evol Microbiol. 2006;56:2729–36.

Ramos DT, da Silva ML, Nossa CW, Alvarez PJ, Corseuil HX. Assessment of microbial communities associated with fermentative–methanogenic biodegradation of aromatic hydrocarbons in groundwater contaminated with a biodiesel blend (B20). Biodegradation. 2014;25:681–91.

Ramos DT, Lazzarin HSC, Alvarez PJ, Vogel TM, Fernandes M, Do Rosario M, et al. Biodiesel presence in the source zone hinders aromatic hydrocarbons attenuation in a B20-contaminated groundwater. J Contam Hydrol. 2016;193:48–53.

Rapant S, Krcmova K. Health risk assessment maps for arsenic groundwater content, application of national geochemical databases. Environ Geochem Health. 2007;29:131e141.

Ray D, Rao RR, Bhoi AV, Biswas AK, Ganguly AK, Sanyal PB. Physico-chemical quality of drinking water in Rohtas district of Bihar. Environ Monit Ass. 2000;61:387.

Renault D, Wallender WW. Nutritional water productivity and diets. Agric Water Manag. 2000;45:275–96.

Ribaudo M, Delgado J, Hansen L, Livingston M, Mosheim R, Williamson J. Nitrogen in agricultural systems: implications for conservation policy. Economic Research Report (ERR Number 127). US Department of Agriculture (USDA), 2011.

Roberts GW, Fortier MP, Sturm BSM, Stagg-Williams SM. Promising pathway for algal biofuels through wastewater cultivation and hydrothermal conversion. Energ Fuel. 2013;27:857–67.

Rosenzweig C, Casassa G, Karoly DJ, Imeson A, Liu C, Menzel A, et al. Assessment of observed changes and responses in natural and managed systems. Climate change 2007: impacts, adaptation and vulnerability. In: Parry ML, Canziani OF, Palutikof JP, van der Linden PJ, Hanson CE, editors. Contribution of working group II to the fourth assessment report of the intergovernmental panel on climate change. Cambridge, UK: Cambridge University Press; 2007. PP. 79–131.

Sampat P. Groundwater shock: the polluting of the world's major freshwater stores. World Watch. 2000;13:10–22.

Schilling KE, Jacobson PJ, Libra RD, Gannon JM, Langel R, Peate DW. Estimating groundwater age in the Cambrian-Ordovician aquifer in Iowa: implications for biofuel production and other water uses. Environ Earth Sci. 2017;76:2–9.

Sesmero JP. Cellulosic biofuels from crop residue and groundwater extraction in the US Plains: the case of Nebraska. J Environ Manag. 2014;144:218–25.

Sobota DJ, Compton JE, Harrison JA. Reactive nitrogen inputs to US lands and waterways: how certain are we about sources and fluxes? Front Ecol Environ. 2013;11:82–90.

Spalding RF, Toso MA, Exner ME, Hattan G, Higgins TM, Sekely AC. Long-term groundwater monitoring results at large, sudden denatured ethanol releases. Ground Water Monit Remed. 2011;31:69–81.

Stafford BP, Capiro NL, Alvarez PJJ, Rixey WG. Pore water characteristics following a release of neat ethanol onto preexisting NAPL. Ground Water Monit Remed. 2009;29:93–104.

Taylor RG, Scanlon B, Döll P, Rodell M, Van Beek R, Wada Y, Longuevergne L, Leblanc M, Famiglietti JS, Edmunds M, Konikow L. Ground water and climate change. Nature Climate Change. 2013;3:322–9.

Tyagi AC. Biofuels and irrigation. Irrig Drain. 2015;64:297–8.

Ullah R, Malik RN, Qadir A. Assessment of groundwater contamination in an industrial city, Sialkot, Pakistan. African J Environ Sci Technol. 2009;3:12.

Vasudevan V, Stratton RW, Pearlson MN, Jersey GR, Beyene AG, Weissman JC, et al. Environmental performance of algal biofuel technology options. Environ Sci Tchnol. 2012;46 (4):2451–9.

Venteris ER, Skaggs RL, Coleman AM, Wigmosta MS. A GIS cost model to assess the availability of freshwater, seawater, and saline groundwater for algal biofuel production in the United States. Environ Sci Tchnol. 2013;47(9):4840–9.

Villemur R, Lanthier M, Beaudet R, Lépine F. The *Desulfitobacterium* genus. FEMS Microbiol Rev. 2006;30:706–33.

Vinson DS, Lundy JR, Dwyer GS, Vengosh A. Implications of carbonate-like geochemical signatures in a sandstone aquifer: radium and strontium isotopes in the Cambrian Jordan aquifer (Minnesota, USA). Chem Geol. 2012;334:280–94.

Ward MH, deKok TM, Levallois P, Brender JD, Gulis G, Nolan BT, et al. Workgroup report: drinking-water nitrate-N and health-recent findings and research needs. Environ Health Perspect. 2005;113:1607–14.

Welch HL, Green CT, Rebich RA, Barlow JR, Hicks MB. Unintended consequences of biofuels production? The effects of large-scale crop conversion on water quality and quantity. U. S. Geological Survey. 2010.

Wilhelm WW, Hess JR, Karlen DL, Johnson JMF, Muth D, Baker JM, et al. Balancing limiting factors and economic drivers for sustainable Midwestern US agricultural residue feedstock supplies. Indus Biotechnol. 2010;6:271–87.

Williams GM, Aitkenhead N. Lessons from Loscoe-the uncontrolled migration of landfill gas. Q J Eng Geol. 1991;24:191–207.

Workocha GA. Impacts of industrial effluents on water body and health of oil producing communities in reverse state. Res J Int Stud. 2011;18:35–40.

Yu S, Freitas JG, Unger AJA, Barker JF, Chatzis J. Simulating the evolution of an ethanol and gasoline source zone within the capillary fringe. J Contam Hydrol. 2009;105:1–17.

Zhang X, Peterson C, Reece D, Haws R, Moller G. Biodegradability of biodiesel in the aquatic environment. Transactions Am Soc Agric Eng. 1998;41:1423–30.

Zhao RF, Chen XP, Zhang FS, Zhang H, Schroder J, Römheld V. Fertilization and nitrogen balance in a wheat–maize rotation system in North China. Agronomy J. 2006;98:938–45.

Chapter 5
Health Concerns Associated with Biofuel Production

Muhammad Arshad, Ijaz Bano, Muhammad Younus, Ammanullah Khan and Abdur Rahman

Abstract Worldwide intensive demand of biofuel as a substitute to fossil fuels has sparked a debate about their advantages especially concerns about human health. Potential health impacts of biofuel are linked to biochemical and chemicals applied in biofuel production processes. Such caustic chemicals are highly hazardous for human health. Other impacts of biofuel come through water pollution; air pollution and use of agrochemicals and pesticides to raise the feedstock. Incomplete burning of sugarcane leaves or residues may results in toxic compounds formation and fine particulates are emitted into atmosphere. The chapter summarizes the basic health effects of biofuel from agriculture cultivation of feedstock to production processes.

Keywords Climate change · Health concerns · Biofuel · Water pollution

5.1 Introduction

Biofuel are normally produced by microbial action that is fermenting sugars from corn grain or sugarcane into ethanol, anaerobic decomposition of organic matter into biogas or through esterification of oil from oil crops into biodiesel. The 2nd generation biofuel that include ethanol from lignocellulose feedstock and, algae-based 3rd generation biofuel are not yet as commercially viable; although initial research demonstrated very promising results in all aspects (Arshad 2010; Dutta et al. 2014; Saidane-Bchir et al. 2016; Lu and Zhang 2016; Losordo et al. 2016; Joelsson et al.

M. Arshad (✉) · I. Bano · M. Younus · A. Khan · A. Rahman
Jhang-Campus, University of Veterinary and Animal Sciences, Lahore, Pakistan
e-mail: muhammad.arshad@uvas.edu.pk

M. Younus
e-mail: younusrana@uvas.edu.pk

A. Khan
e-mail: amanullah.khan@uvas.edu.pk

A. Rahman
e-mail: abdrrahman@uvas.edu.pk

M. Arshad (ed.), *Perspectives on Water Usage for Biofuels Production*,
https://doi.org/10.1007/978-3-319-66408-8_5

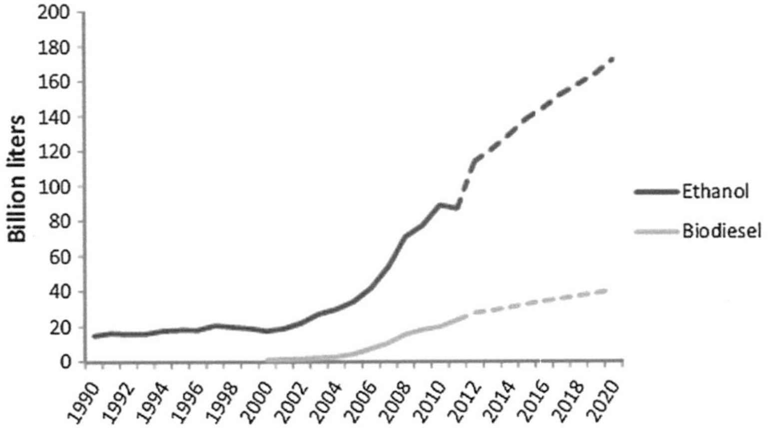

Fig. 5.1 Global production of ethanol and biodiesel (Scovronick and Wilkinson 2014)

2016; Skuland et al. 2017; Bhuiya et al. 2016; Mahdy et al. 2017; Sturmer 2017). The rapid expansion of the biofuel production in last two decades have forced many researchers (Galdos et al. 2013; Ridley et al. 2012; Swanson et al. 2007; Miraglia 2007; McCormick 2007; Poeschl et al. 2012; Hill et al. 2009; Morris et al. 2003) to explore the health impacts of these biofuel. In view of that, the present chapter is focused on health impact of biofuel.

More or less 90% of the global bioethanol is produced in United States (US) and Brazil (Chum et al. 2014). US utilizes corn and while Brazil uses sugarcane as the main feedstock for bioethanol production. Figure 5.1 shows global production of ethanol and biodiesel. Environmental effects such like; biodiversity loss, deforestation, erosion of soils and contamination of water assets are on rise, with the worldwide increased demand of fuel ethanol production. Sugarcane is better grown in tropical hot areas, where presently tropical forests and biodiversity hotspots are present to sustain natural ecosystem. Such regions are going to use for growing of biofuel feedstock in the future (Solomon and Bailis 2014).

Environmental impacts related to biofuel production and its uses are very hard to be figured out. Low air pollution is expected from ethanol and biodiesel as compared to fossil based fuels. Major health problems associated with the biofuel production industry are due to watershed problems and groundwater contamination. Agrochemicals and fertilizers use is the major cause of such issues. In this chapter, health related issues of biofuel have been covered.

5.2 Health Concerns During Production Processes

Biochemical and chemicals are applied in production processes of biofuel. All of these are also hazardous to human health. Sulfuric acid is used to adjust the pH of fermentation media during commercial production of ethanol (Arshad et al. 2008,

2011, 2014b, 2017). Sodium hydroxide is applied for sterilization and cleaning in process, while ammonia is used as nitrogen source. All the three chemicals are caustic in nature. More over yeasts, enzymes and antibiotics are used.

The ethanol itself has potential to cause skin and eye irritation (Bevan et al. 2009). Likewise, production of biodiesel takes in reacting lipids with an alcohol (Arshad et al. 2014a). Generally methanol is used that causes serious health issues. Caustic chemicals, sodium or potassium hydroxide are employed as catalyst to yield biodiesel and glycerol (Swanson et al. 2007). In biogas production process risks of air pollution, explosions and fires are involved. Release of H_2S is also very hazardous for human health.

5.3 Health Concerns Due to Water Pollution

Inhalation, dermal contact or consumption, can bring water and soil pollutants in contact to humans. Infectious, pathogenic or chemical nature of pollutant along with exposure time determines the health impacts. Such impacts range from skin irritation to carcinogenic level (Scovronick and Wilkinson 2014). Not only water quantity but also the water quality is affected through the cultivation of biofuel crops. Grave environmental and human health effects are linked with sugar industries comprising water pollution due to application of fertilizers, use of various agrochemicals to protect the sugarcane crop from pests, throwing away of unsatisfactorily treated wastewater from the alcohol and sugar processing plants, soil erosion, among others. Biofuel life-cycle and major pathways linked with health concerns are depicted in Fig. 5.2.

5.3.1 Generation of Wastewater

Major liquid effluent of bioethanol production is the stillage/vinasse and the wastewater generated through cleaning operations. Disposal of stillage is the most significant as it is produced in huge quantity, almost 10–12 l for production of 1 L ethanol (Palacios-Bereche et al. 2014). Moreover, it has high biochemical oxygen demand and chemical oxygen demand) with the acidic pH (4–5) (Janke et al. 2016). In Pakistan, most of distilleries store stillage into open ponds for sun drying (Khuhawar et al. 2011; Mahar et al. 2012, 2013), so it is leached to the underground water and contaminates it. Even the color of water is changed to brownish. Various skin diseases are found in the nearby areas.

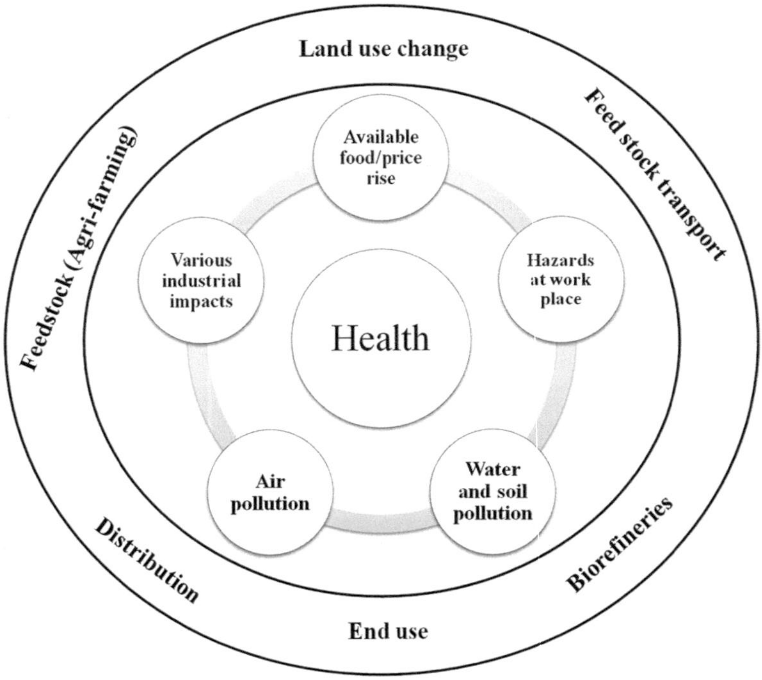

Fig. 5.2 Schematic depiction of the biofuel life-cycle and major pathways linked with health concerns

5.3.2 Applications of Agrochemicals and Pesticides

Comparatively low nitrogen input is required for soybean biodiesel matched with ethanol from corn or switch grass (Dominguez-Faus et al. 2009). But soybeans had the highest pesticide use and switch grass the lowest. Biofuel spills during storage and distribution is another source of contamination, particularly as ethanol increases the risk of storage tank corrosion (Niven 2005).

Aquifer keeps distinct water quality that is disturbed by the nitrates; potassium; chloride; calcium and magnesium originated from above mentioned pollutants. Application of inorganic fertilizers and pesticides (Bava 2010) on sugarcane crop is a normal practice. When concentration of such ions goes above permissible limits in drinking water, a serious threat to human health seems. Partly filtered and digested effluents from the industrial operations are also applied to the sugarcane fields as a fertilizer. Such biogas waste, denominated as vinasse or stillage, is nutrient rich; eutrophication is caused and pollutants runoff to water bodies.

5.3.3 *Harvesting of Sugarcane*

Manual sugarcane harvesting is very laborious work and may associate with minor injuries. Excess work or physical exertion in the sugarcane field may cause fatigue and exhaustion (De-Souza et al. 1998; Smeets et al. 2009). Some secondary health issues arise in neighboring societies that use agrochemical contaminated groundwater for drinking purposes. Solar irradiation, respiratory issues are major among the environmental factors. It is very difficult to connect human health and pesticides directly. Risk factors mostly include the carcinogenic characteristics of agrochemicals (McKelvey et al. 2004).

5.4 Health Concerns Due to Burning

Sugarcane crop is normally burned in advance to harvest it to ease the cutting of the cane stalks. Burning may repel the poisonous animals, snakes and spiders but it damages the tissue cells of the cane stem, making it susceptible to microbes that can cause diseases, may destroy organic matter and damage the soil structure due to increased drying with the increased risks regarding erosion of soils. There are unsafe atmospheric emissions such as CO, CH_4, non-methane organic compounds and particulate matter (Smeets et al. 2009; Arshad and Ahmed 2016), their health concerns have been presented in Table 5.1.

Respiratory issue can rise due to burning of crops and the sugarcane leaves. The existence of particulate matter and gases may cause cardiovascular diseases and lung cancer in workers. Chances of acid rain are increased (Goldemberg et al. 2008; Cancado et al. 2006). Sometimes, incomplete combustion of biomass may results in toxic compounds formation. Polycyclic aromatic hydrocarbons (PAHs), methane and fine particulates are emitted. Consequently, edible products coming from sugarcane factories may keep some traces of pesticides, their metabolic products and equally the chance of PAHs.

Table 5.1 Atmospheric emission due to burning and health concerns

Probable Gas	Disease	References
CO	Respiratory and cardiovascular issues, poisoning	Ernst and Zibrak (1998), Weaver (2009)
NO_2	Respiratory issues	Kampa and Castanas (2008), Ghozikali et al. (2015)
O_3	Eye irritation Respiratory problems (inflammatory reaction of the respiratory system)	Ghozikali et al. (2015)
Pb	Cumulative toxic effect Anemia and brain tissue destruction	Juberg et al. (1997)
SO_2	Respiratory problems, eye irritation and cardiovascular problem	Ghozikali et al. (2015)

5.5 Risk of Cancer

The environment is at serious risk due to overuse of pesticides on the sugarcane crop. Some pesticides have been identified as endocrine disruptors. Endocrine system of mammals (birth defects and infertility), diabetes, cancer and even changes in behavior are caused by estrogenic activity of agrochemicals especially pesticides. Chlorine derivative pesticides can cause cancer (Vieira et al. 2008). The nitrates, soluble in water, a macronutrient of the plants, originated from nitrogen, provided by inorganic fertilizer and animal manure, are the primary form of nitrogen, present in the ground water of agricultural lands as it can easily pass through soil to groundwater table.

Nitrates are very stable and can persist in groundwater for decades, so may accumulate to high level and promote risk of many type of cancer in human stomach and colon (Ward et al. 2005; Irigaray et al. 2007). Newborns, especially below six months age are more vulnerable. Pesticides contamination can be associated with breast cancer incidence (O'Leary et al. 2004; Mittal et al. 2014). Alike study was conducted by Brody et al. (2006) to diagnose the cancer in women of the peninsula of Cape Cod. A correlation between the etiology of cancer and the exposure to pesticides contaminated groundwater was established.

5.6 Conclusion

Swift growth of biofuel may impact human health through a range of pathways. Humans can be exposed to hazards chemicals use in production process, water pollution and air pollution caused by the biofuel. Effects of air pollution and burning are highly considerable and have been relatively well studied in many respects. Health impact of first generation biofuel especially ethanol from sugarcane have been much explored. Future generation biofuel are likely to become more competitive in future with less health impacts. The cellulose based biofuel have the potential to lessen the major health issues linked with first-generation biofuel. Feedstock crops of future biofuel may require less water, fertilizers and pesticides as compared to biofuel of first generation. Emissions associated with production process will seem to be less. Following precautions for future biofuel should be adopted.

1. Use of biochemical and chemical reagents in biofuel processes must be monitored.
2. Conduct of epidemiological studies to understand the health risk associated with seasonal exposure to biomass burning.
3. Heath impact of air emissions at all stages in the lifecycle of biofuel must be considered.

4. Future biofuel policies must be designed keeping in view the health impact of biofuel.
5. The products coming from biofuel refineries may keep some traces of pesticides or their metabolic products. It may be analyzed carefully.

References

Arshad M, Khan ZM, Shah FA, Rajoka MI. Optimization of process variables for minimization of byproduct formation during fermentation of blackstrap molasses to ethanol at industrial scale. Lett Appl Microbiol. 2008;47:410–4.

Arshad M. Bioethanol: A sustainable and environment friendly solution for Pakistan. A Scientific J COMSATS–Sci. Vision. 2010;16–7.

Arshad M, Zia MA, Asghar M, Bhatti H. Improving bio-ethanol yield: Using virginiamycin and sodium flouride at a Pakistani distillery. Afr J Biotechnol. 2011;10:11071.

Arshad M, Adil M, Sikandar A, Hussain T. Exploitation of meat industry by-products for biodiesel production: Pakistan's perspective. Pakistan J Life Soc Sci. 2014a;12:120–5.

Arshad M, Ahmed S, Zia MA, Rajoka MI. Kinetics and thermodynamics of ethanol production by *Saccharomyces cerevisiae* MLD10 using molasses. Appl Biochem Biotechnol. 2014b;172:2455–64.

Arshad M, Ahmed S. Cogeneration through bagasse: a renewable strategy to meet the future energy needs. Renew Sust Energ Rev. 2016;54:732–7.

Arshad M, Hussain T, Iqbal M, Abbas M. Enhanced ethanol production at commercial scale from molasses using high gravity technology by mutant *S. cerevisiae*. Brazilian J Microbiol. 2017. doi:10.1016/j.bjm.2017.02.003.

Bava SC. Alimentos Contaminados. Le Monde Diplomatique Brasil 2010; ed.33.

Bevan RJ, Slack RJ, Holmes P, Levy SL. An assessment of potential cancer risk following occupational exposure to ethanol. J Toxicol Environ Health B Crit Rev. 2009;12:188–205.

Bhuiya MMK, Rasul MG, Khan MMK, Ashwath N, Azad AK. Prospects of 2nd generation biodiesel as a sustainable fuel-Part: 1 selection of feedstocks, oil extraction techniques and conversion technologies. Renew Sust Energ Rev. 2016;55:1109–28.

Brody JG, Aschengrau A, McKelvey W, Swartz CH, Kennedy T, Rudel RA. Breast cancer risk and drinking water contaminated by wastewater: a case control study. Environ Health: A Global Access Sci Source. 2006;5:28.

Cançado JE, Saldiva PH, Pereira LA, Lara LB, Artaxo P, Martinelli LA, Arbex MA, Zanobetti A, Braga AL. The impact of sugar cane-burning emissions on the respiratory system of children and the elderly. Environ Health Perspect. 2006:725–9.

Chum HL, Warner E, Seabra JE, Macedo IC. A comparison of commercial ethanol production systems from Brazilian sugarcane and US corn. Biofuels, Bioprod Biorefin. 2014;8:205–23.

De-Souza DA, Marchesan WG, Greene LJ. Epidemiological data and mortality rate of patients hospitalized with burns in Brazil. Burns. 1998;24:433–8.

Dominguez-Faus, Rosa, Susan E. Powers, Joel G. Burken, Pedro J. Alvarez. The water footprint of biofuels: a drink or drive issue? 2009:3005–10.

Dutta K, Daverey A, Lin JG. Evolution retrospective for alternative fuels: first to fourth generation. Renew Energ. 2014;69:114–22.

Ernst A, Zibrak JD. Carbon monoxide poisoning. New England J Med. 1998;339:1603–8.

Galdos M, Cavalett O, Seabra JE, Nogueira LAH, Bonomi A. Trends in global warming and human health impacts related to Brazilian sugarcane ethanol production considering black carbon emissions. Appl Energ. 2013;104:576–82.

Ghozikali MG, Mosaferi M, Safari GH, Jaafari J. Effect of exposure to O_3, NO_2, and SO2 on chronic obstructive pulmonary disease hospitalizations in Tabriz, Iran. Environ Sci Pollut Res. 2015;22:2817–23.

Goldemberg J, Coelho ST, Guardabassi P. The sustainability of ethanol production from sugarcane. Energ Policy. 2008;36:2086–97.

Hill J, Polasky S, Nelson E, Tilman D, Huo H, Ludwig L, et al. Climate change and health costs of air emissions from biofuels and gasoline. Proc Natl Acad Sci. 2009;106:02077–82.

Irigaray P, Newby JA, Clapp R, Hardell L, Howard V, Montagnier L, et al. Life style-related factors and environmental agents causing cancer: an overview. Biomed Pharmacotherap. 2007;61:640–58.

Janke L, Leite AF, Nikolausz M, Radetski CM, Nelles M, Stinner W. Comparison of start-up strategies and process performance during semi-continuous anaerobic digestion of sugarcane filter cake co-digested with bagasse. Waste Manage. 2016;48:199–208.

Joelsson E, Erdei B, Galbe M, Wallberg O. Techno-economic evaluation of integrated first-and second-generation ethanol production from grain and straw. Biotechnol Biofuel. 2016;9:1.

Juberg DR, Kleiman CF, Kwon SC. Position paper of the American Council on Science and Health: lead and human health. Ecotoxicol Environ Safety. 1997;38:162–80.

Kampa M, Castanas E. Human health effects of air pollution. Environ Pollut. 2008;151:362–7.

Khuhawar MY, Baloch MA, Jahangir TM, Mahar MT, Majidano SA. Impacts of evaporation ponds of ethanol distillery spent wash on underground water. Pak J Chem. 2011;1(1):10–8.

Losordo Z, McBride J, Rooyen JV, Wenger K, Willies D, Froehlich A, et al. Cost competitive second-generation ethanol production from hemicellulose in a Brazilian sugarcane biorefinery. Biofuel Bioprod Biorefin. 2016;10:589–602.

Lu D, Zhang XJ. Biogas production from anaerobic codigestion of microalgae and septic sludge. J Environ Eng. 2016;142:04016049.

Mahar MT, Khuhawar MY, Baloch MA, Jahangir TM. Effects of spent wash of ethanol industry on groundwater: a case study of Rahimyar Khan district, Pakistan. J Environ Sci Water Resour. 2012;1:85–94.

Mahar MT, Khuhawar MY, Jahangir TM, Baloch MA. Health risk assessment of heavy metals in groundwater: the effect of evaporation ponds of distillery spent wash. J Environ Sci Eng. 2013;2(3A):166.

Mahdy A, Fotidis IA, Mancini E, Ballesteros M, González-Fernández C, Angelidaki I. Ammonia tolerant inocula provide a good base for anaerobic digestion of microalgae in third generation biogas process. Bioresour Technol. 2017;225:272–8.

McCormick RL. The impact of biodiesel on pollutant emissions and public health. Inhal Toxicol. 2007;19:1033–9.

McKelvey W, Brody JG, Aschengrau A, Swartz CH. Association between residence on Cape Cod, Massachusetts, and breast cancer. Ann Epidemiol. 2004;14:89–94.

Miraglia SGEK. Health, environmental, and economic costs from the use of a stabilized diesel/ethanol mixture in the city of Sa˜o Paulo, Brazil. Cad.Sau′ de Pu′ blica. 2007;23:S559–S569.

Mittal S, Kaur G, Vishwakarma GS. Effects of environmental pesticides on the health of rural communities in the Malwa Region of Punjab, India: a review. Human Ecol Risk Ass: An Int J. 2014;20:366–87.

Morris R, Pollack A, Mansell G, Lindhjem C, Jia Y, Wilson G. Impact of biodiesel fuels on air quality and human health: summary report. Golden, Colorado: National Renewable Energy Laboratory; 2003.

Niven RK. Ethanol in gasoline: environmental impacts and sustainability review article. Renew Sust Energ Rev. 2005;9:535–55.

O'Leary ES, Vena JE, Freudenheim JL, Brasure J. Pesticide exposure and risk of breast cancer: a nested case-control study of residentially stable women living on Long Island. Environ Res. 2004;94:134–44.

Palacios-Bereche R, Ensinas A, Modesto M, Nebra SA. New alternatives for the fermentation process in the ethanol production from sugarcane: extractive and low temperature fermentation. Energy. 2014;70:595–604.

Poeschl M, Ward S, Owende P. Environmental impacts of biogas deployment–Part I: life cycle inventory for evaluation of production process emissions to air. J Clean Prod. 2012;24:168–83.

Ridley CE, Clark CM, Leduc SD, Bierwagen BG, Lin BB, Mehl A, et al. Biofuels: network analysis of the literature revealskey environmental and economic unknowns. Environ Sci Technol. 2012;46:1309–15.

Saïdane-Bchir F, El Falleh A, Ghabbarou E, Hamdi M. 3rd generation bioethanol production from microalgae isolated from slaughterhouse wastewater. Waste Biomass Valor. 2016;7:1041–6.

Scovronick N, Wilkinson P. Health impacts of liquid biofuel production and use: a review. Glob Environ Change. 2014;24:155–64.

Skuland TS, Refsnes M, Magnusson P, Oczkowski M, Gromadzka-Ostrowska J, Kruszewski M, et al. Pro inflammatory effects of diesel exhaust particles from moderate blend concentrations of 1st and 2nd generation biodiesel in BEAS-2B bronchial epithelial cells-the fuel health project. Environ Toxicol Phar. 2017;52:138–42.

Smeets EM, Bouwman LF, Stehfest E, Vuuren V, Detlef P, Posthuma A. Contribution of N2O to the greenhouse gas balance of first-generation biofuels. Glob Change Biol. 2009;15:1–23.

Solomon BD, Bailis R, editors. Sustainable development of biofuels in Latin America and the Caribbean. New York: Springer; 2014.

Stürmer B. Feedstock change at biogas plants–Impact on production costs. Biomass Bioenerg. 2017;98:228–35.

Swanson KJ, Madden MC, Ghio AJ. Biodiesel exhaust: the need for health effects research. Environ Health Perspect. 2007;115:496–9.

Vieira VM, Webster TF, Weinberg JM, Aschengrau A. Spatial-temporal analysis of breast cancer in upper Cape-Cod, MA. Int J Health Geogr. 2008;7:46.

Ward MH, DeKok TM, Levallois P, Brender J, Gulis G, Nolan BT, VanDerslice J. Workgroup report: drinking-water nitrate and health-recent findings and research needs. Environ Health Perspect. 2005:1607–14.

Weaver LK. Carbon monoxide poisoning. New England J Med. 2009;360:1217–25.

Chapter 6
Future Biofuel Production and Water Usage

Muhammad Arshad and Mazhar Abbas

Abstract Biofuel in particular, together with the rising demands for food, have the highest prospects for an increase in agricultural water withdrawals. The water-biofuel relationship is being recognized as backbone of the factors fundamental for the future sustainable supply of water and biofuel. A better understanding of the subject is essential to adopt superior technologies that may improve use of water for biofuel production in efficient way. This chapter presents prospective and future trends of the water-biofuel relationship and impacts of additional water usage in future increased biofuel production. The importance of technological innovation to save water and future impacts on water quantity and especially on water quality will be assessed in terms of safe keeping the environment. The obligation of reusing wastewater and application of undiluted wastewater to grow feedstock for biofuel to save freshwater resources will be analyzed.

Keywords Climate changes · Future water availability · Biofuel production · Fresh water assessment

6.1 Introduction

Biofuel offer better answers to present world energy needs and economic crises, both as a sustainable energy source and through promoting economic development, especially in rural areas of developing countries (Joly et al. 2015; Arshad 2010; 2011; 2014a; 2017). Biofuel are regarded most promising renewable substitutes of fossil fuels to meet the aim to pull down CO_2 emissions (Farrell et al. 2006; Ragauskas et al. 2006; Arshad et al. 2008, 2014b). Share of biofuels for road transport in various countries has been presented in Fig. 6.1.

M. Arshad (✉) · M. Abbas
Jhang-Campus, University of Veterinary and Animal Sciences, Lahore, Pakistan
e-mail: muhammad.arshad@uvas.edu.pk

M. Abbas
e-mail: Mazhar.abbas@uvas.edu.pk

M. Arshad (ed.), *Perspectives on Water Usage for Biofuels Production*,
https://doi.org/10.1007/978-3-319-66408-8_6

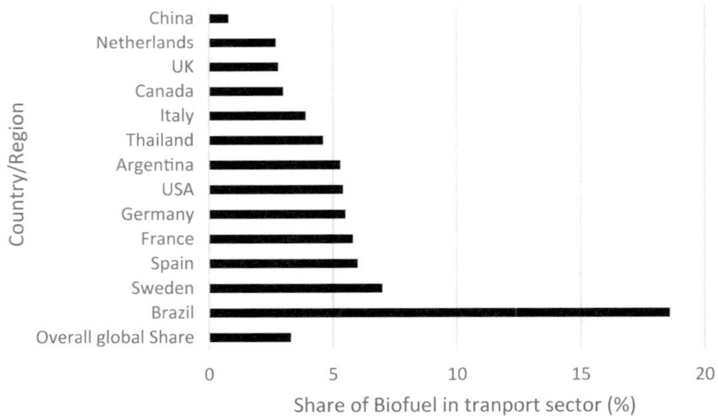

Fig. 6.1 Share of biofuels for road transport in 2011 (Raboni et al. 2015)

The expansion of the feedstock production for biofuel has been controversial due to potential adverse side effects on natural ecosystems and the services they provide (Gasparatos et al. 2011). Globally land and water hungry nature of biofuel feedstock has been matter of concern now. Not good choices in selection of feed stocks and agricultural practices may emasculate environmental goals of biofuel and resource sustainability. Certainly major water resources are used to irrigate the agriculture farming, employed to raise food. But other sector such as energy, electricity and fuel (biofuel) production also need increasing amounts of water (Macknick et al. 2012; Arshad and Ahmed 2016). In certain areas, this trend has forced to a struggle between various water uses (Cosgrove and Rijsberman 2014). Simultaneously the climatic variations can also shrink the availability of water with drop in the quality (Jiménez Cisneros et al. 2014). From now, the energy and transport fuel sources we select, can speed up the rising water requirements or offset such needs in future. Water is the vital natural resource. Its linkage between water use, energy requirement and food production is multifaceted, as the changes in the need of one resource in one sector can change its availability and that of another resource in another sector and vice versa.

Up till now, biofuel production cost comparison to fossil fuels has been remained focused with large scale agribusinesses of the crops investing heavy energy inputs, high pesticide, fertilizer, and water use. The need of the day is to calculate the ecological footprint caused by large-scale cultivation of a given biofuel energy crop, that includes different modes of processing the feedstock into a liquid fuel. Numerous factors are involved in ecological footprints of biofuel. Over all energy balance with energy efficiency in the complete life cycle of biofuel and generation from fuels produced per hectare impacts the whole ecological footprint. These factors impacts on the land required for growth of enough masses of biofuel to substitute fossil based fuels up to significant level. Carbon intensity of biofuels resulting from different feedstock and technology, compared to traditional fossil fuels has been shown in Fig. 6.2.

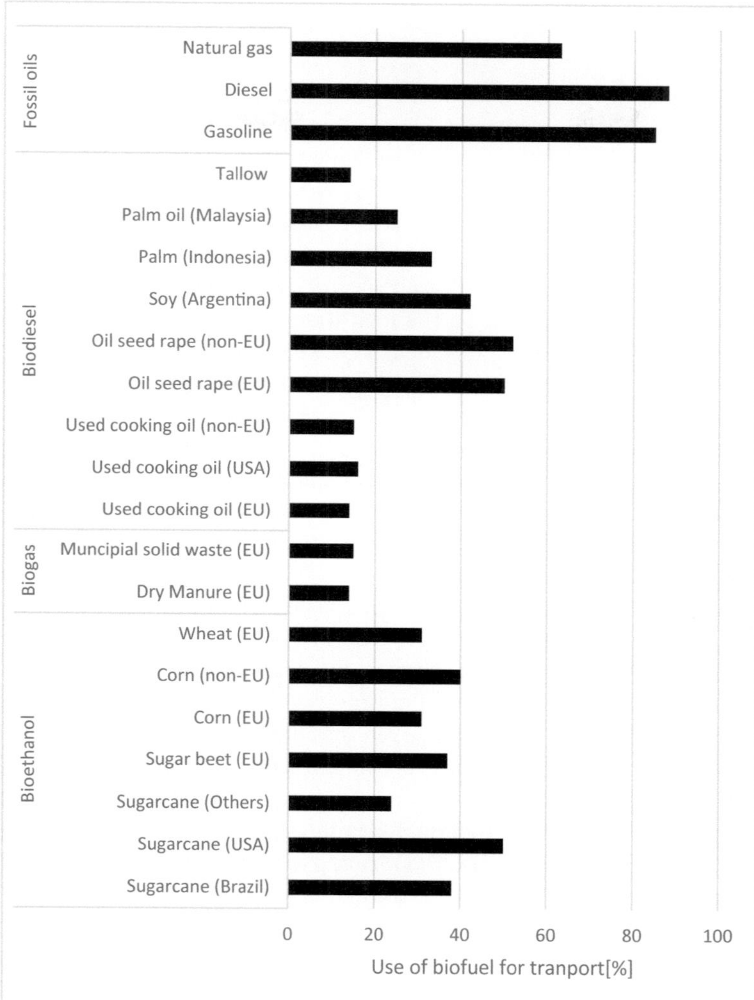

Fig. 6.2 Carbon intensity of biofuels resulting from different feedstock and technology, compared to traditional fossil fuels (Raboni et al. 2015)

The amount of water, pesticide and fertilizer utilized, energy required in farming of the feedstock and greenhouse gas emissions over the life cycle of the product are included in estimation of ecological footprint of a biofuel. Groom et al. (2008) estimated these variables for presently leading biofuel feedstock. This chapter deals largely with the effects of future expansion in feedstock production of biofuel on water, land and ecosystem. It has been also focused on proactive solutions to avoid or minimize potential adverse impacts.

6.2 Vision 2050 and Water Requirements

Climatic changes will create more doubts in the accessibility of freshwater ensuring drinking quality. It is firm scientific believe that human induced emissions of CO_2 are assembling swiftly in the atmosphere. Now it is the fact that the global water resources will be under pressure in 2050 as world population will reach up to 9 billion (Bakker and Morinville 2013). Water need is foreseen to upsurge by 55% globally between 2000 and 2050, demand in certain industries, such as manufacturing will be increased by 400% and electricity production by 140% (OECD 2012). Sources of fresh water as part of total water on earth have been presented in Fig. 6.3. The world may see a 40% deficit between forecasted water requirements and available supplies (2030 Water Resources Group 2009). The core associations among water security and the world's rapidly growing needs for food and fuel make this challenge more complicated.

As the 60% increased food production is required by 2050 to fulfill the growing population's feed demand (Alexandratos and Bruinsma 2012). Simultaneously, climate change will impact agricultural farming, and biofuel demands will contest with food for the same water and land use. In effect, food, biofuel, fiber and ecosystems are all going to screech for more water. The water biofuel competition related to climate changes has been depicted in Fig. 6.4.

Almost 165 billion m^3 of wastewater is collected globally and treated, but just 2% is reused. It shows the potential of reusing wastewater as the same water can be used several times before being discharged into the natural environment. All types of water cannot be used for recycling, so water quality standards should be adapted to the desired end use. Globally almost 1.3 billion tons of eatable food is wasted every year (Conforti 2011). Similarly 7.5 million tons of food waste is guided to landfill in Australia (Mason et al. 2011). Along with the problems associated with waste disposal, we are consuming vast amounts of water as embedded in these losses (Chartres and Sood 2013).

Worldwide water availability may be reduced and limited water resources for agriculture, over the coming decades will be available (Hanjra and Qureshi 2010). Chartres and Sood (2013) analyzed the water demand for food production until 2050 using the WATERSIM model. An increase in global water demand for agriculture from 2400 km^3/yr in 2010 to between 3820 and 7230 km^3/yr in 2050 were forecasted. An increase from 1425 km^3/yr irrigation (blue) water demand for crop production in 2000 to 1785 km^3/yr in 2050 in their baseline scenario was forecasted by Sulser et al. (2010) using IFPRI's IMPACT model. Many water availability analysis have been performed in previous years (Khan et al. 2009; Perrone et al. 2011; Scott et al. 2011; Hardy et al. 2012; Larson 2013; Lele et al. 2013; Rasul 2014). All were agreed that water availability is expected to decline due to rising demands and simultaneous adverse ecological changes.

The energy in the form of scaling up biofuel production in the future need water and land requirements, with their potential adverse effects on food security and water availability has been summarized by (Dominquez-Faus et al. 2009; Yang

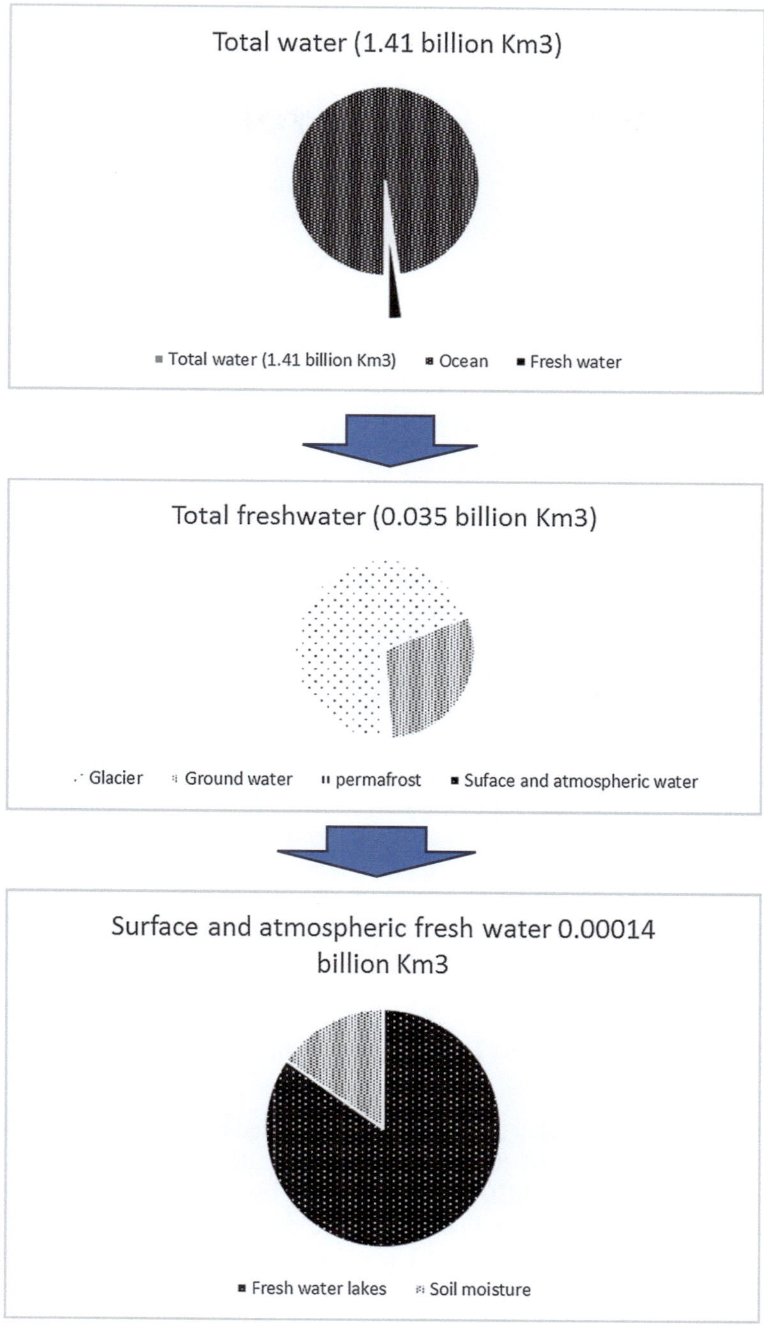

Fig. 6.3 Sources of freshwater as part of total water on earth

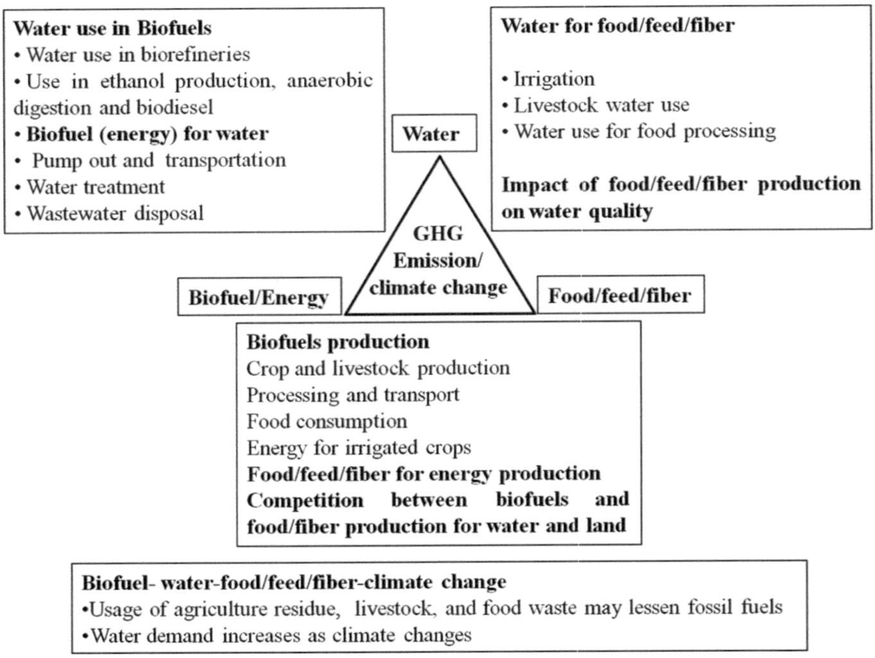

Fig. 6.4 The water biofuel compitition related to climate changes

et al. 2009; Fingerman et al. 2010).Water needed for maize as energy crop in the US, and its harmful effects regarding water availability and environmental health were analyzed by Dominquez-Faus et al. (2009). Fingerman et al. (2010) reported that huge quantity of water (5100 L/L) will be needed for ethanol production. The water and land necessities for biofuel production in China were summarized by Yang et al. (2009).

6.2.1 Increase in First-Generation Biofuel

The water feeding for biofuel (energy) can be increased up to 2012 from 74 km^3/yr in 2050 (97% will be used for growing biomass), if worldwide per capita fuel requirement would reach OECD levels and 7% of the need should be met by first generation biofuel. First generation biofuel are produced in every corner of the world from common energy crops and quantity of water may equal the amount of water required for increased food supply.

An analysis of global biofuel policies and their consequences on water requirements in agricultural sector worldwide was performed. As China and India, are rapidly rising economies and will require more water from limited water resources. It may lead to a tough resource rivalry in the future, if biofuel were

Table 6.1 EU biofuel production and use for transport in 2012 (keto $^{y-1}$) (Raboni et al. 2015)

Country	Bioethanol		Biodiesel	
	Production	Use of transport	Production	Use of transport
Germany	387	403	2861	2191
France	600	209	1910	2268
Spain	191	101	925	1899
Italy	75	40	706	1264
Poland	106	77	370	669
Austria	108	34	289	390
Other member states	952	538	2509	2980
Total	2418	1401	9570	11,661

employed as major transport fuels. Worldwide irrigation water withdrawals for bioethanol production may reach to 128.4 in 2030 from 30.6 km^3/yr in 2005.

To calculate the water utilization in food as well as for first-generation biofuel production, data from the global Water Footprint (WFP) Network was utilized by Damerau et al. (2016). The volume of fresh water taken to yield a product, including the water quantity utilized and polluted in the all the steps of the supply chain are defined as Water Footprint. Gerbens-Leenes et al. (2008) listed blue and green water consumption for production of bioenergy. Fresh surface and groundwater is blue water while the gray water is the quantity of water needed to dilute pollutants. Water coming precipitation on land that does not run off or re-charge the groundwater but is stored in the soil or temporarily stays on top of the soil or vegetation is called green water (Ridoutt and Pfister 2013; Sulser et al. 2010).

First generation biofuel can have large negative ecological impacts, not only with regard to water (Creutzig et al. 2014). So the European Union revised their biofuel targets until 2020, limiting first-generation biofuel to a share of 7% (Eggert and Greaker 2014). EU biofuel production and use for transport in 2012 has been shown in Table 6.1. An increase in first-generation biofuel can easily lead to huge needs of extra water and demand for cropland would also rise, which might lead to additional competition for landwith food production (Rathmann et al. 2010).

6.3 Future Impacts on Ecosystems

Ecosystems are affected due to agricultural activities, either to raise food and fiber or biofuel. One of major impact related to biofuel is the land-use change. The required land to raise feedstock for biofuel production may come rightly after clearing new land, but mostly as a consequence of replacing the one crop with another (Hertel et al. 2010; Searchinger et al. 2008). The ecological effects of biofuel are interceded the impact on land, water, air complexly connected with the economics of worldwide agricultural markets. Ecology is an emergent part of biology that has to lead us towards the sustainable production of biofuel in a cost effective and environmentally safe way. It may be employed to produce large quantities of feedstock keeping the desired chemicals composition. If feedstock for biofuel are raised through sustainable means then biofuel can be sustainable source of energy and may be promoted with smallest ecological footprint. The Ecological Footprint measures the amount of biologically productive land and water area required to produce all the resources that an individual, a population, or an activity consumes, considering also the absorption of residues they generate. This can be compared to the biocapacity, the amount of productive area that is available to generate these resources and to absorb the residues (Wang 2005).

6.3.1 Land Use Change

According to Fargione et al. (2010), the land needed for biofuel production can be easily estimated by dividing the biofuel quantity with the conversion efficiency multiplied by crop yield and un-harvested correction. Few biofuel also generate byproducts and coproducts that can substitute many other products in the market place, decreasing the net amount of food displaced. Resultantly, the quantity of land needed to produce such biofuel can be lesser. Therefore the impact of co-product/ byproduct must be incorporated. It is estimated that 15.9 million ha were used to produce ethanol and 17.4 million ha were cultivated for biodiesel production in 2008. It approaches to 2.2% of worldwide cropland.

Assuming biofuel expand by 170% in 2020, as under a business as usual scenario (Fargione et al. 2010), cropland required for biofuel production would be 72–82 million ha if biofuel production efficiencies (that is, crop yields and conversion efficiencies) increased by 10–25%. Our estimates of the land required to produce biofuel do not include co-products effects due to lack of data. Research on coproduct effects could help guide biofuel producers toward processes and coproducts that reduce the amount of new land required for biofuel production.

6.4 Possible Solution

6.4.1 Polyculture Versus Monoculture

The type of agriculture farming growing more than one crop at a time on the same space is called polyculture. It provides crop diversity to maintain the natural ecosystems. Biofuel feedstock needing some inputs, consuming innate species or that emphasize perennial species, mostly in polyculture can be better biodiversity friendly as compared to energy intensive monocultured yearly crops. The value of biofuel crop in terms of biodiversity can be increase through polyculture technique.

It may decrease the pest and soil fertility issues (Tilman et al. 2006). Conservative biologists may give better input by estimating the biodiversity costs and advantages in cultivation of major portion of land through polyculture technique.

6.4.2 Less Input Feedstock

One possibility to reduce the water requirements for first-generation biofuel production would be a shift towards energy crops that show lower water demands but are currently less often used. However, a general restriction of first-generation biofuel as well as the deployment of freshwater-cooled thermal energy technologies in the future would also limit the additional water (and land) demand in the energy sector, an increase that could be more than offset by changes in the food sector. Due to this potential trade-off, an overall increase in water demand in both sectors is not necessarily an unavoidable trend. Our results provide valuable new insights and information for integrated natural resource management and policy, in particular with respect to biofuel targets. Mitigation measures as discussed in previous studies can further improve water efficiency, especially in regions where water availability might decline over the next decades as a consequence of climate change and other potential ecological changes.

Whether the biofuel crop can give better energy yields per hectare under less input techniques; switchgrass that is grown with much less fertilizer inputs than other crops, especially corn can be the better answer (Graham et al. 1995; Parrish and Fike 2005). The switchgrass has been investigated more comprehensively as compare to any other feedstock. Therefore it can well lead towards advances in best farming practices with high yield and energy extraction (Parrish and Fike 2005). Switchgrass seizes carbon below ground, resulting in a negative greenhouse gas balance (Adler et al. 2007). Other perennial prairie grasses that can serve as biodiversity friendly feedstock must be explored. Wood, crop residues and other perennial species can be ecologically better than grain and grass feedstock for biofuel production. Municipal or industrial wastewater irrigated, poplar and willow plant can also be better feedstock as these can decrease waste streams while achieving inputs needed for high yields (Powlson et al. 2005). Aptness of woody biomass also based on either the native species are used and plants were grown in sustainable style. Conversion of forests to tree plantations with short rotation tree species can be most appropriate for biofuel production, especially Populus species (Popular) and Salix species (willows). The tree energy crops can enhance biodiversity. If biofuel production from woody biomass becomes profitable, it might serve to motivate land restoration and to avoid conversion of native habitats. Meeting current global demand for petroleum via current-generation biofuel would require a doubling of the human share of net primary productivity, which would threaten species and habitats with extinction and sharply decrease global food security (Junginger et al. 2006). Thus, many look to high-efficiency extraction of hydrocarbons from lignocellulosic biomass as a necessary precondition to successful use of biofuel (EEA 2006). Still there exist some practical difficulties in conversion of lignocellulosic biomass to ethanol for production of fuel ethanol at commercial scales. Thus far the conditions are not well defined under which biofuel derived from woody biomass are biodiversity-friendly.

Lal (2006) stressed that crop residues mostly play their role in maintenance of soil fertility and in reduction of soil erosion from rain and wind, and suppress weed growth. Therefore the use of agriculture residue in biofuel production may affect the agriculture. Health of the soils is gauge of crop yields grown for energy or for food. Biofuel crops differ in soil fertility and fertilizer requirements, and in types of soil management or conservation practices compatible with high yields. Switchgrass is better as it takes nitrogen efficiently from soils as compared many other species as corn and other grasses (Parrish and Fike 2005). Very less inputs of fertilizer are needed when mixed native prairie grasses were grown on degraded soils (Tilman et al. 2006). In contrast, corn and soy is cultivated under significant fertilizer quantity (Griffing et al. 2014). That results in major nitrogen overspills into surrounding and distant waterways. Thus greenhouse gas emissions (Powlson et al. 2005; Hill et al. 2006) are increased.

Present energy harvesting efficacies compels to grow energy crops on a massive spatial scale to replace even half of U.S. transportation fuel demands. That has huge consequent effects on biodiversity. Over 20–50% portion of the land in terrestrial biomes has been reserved for food production for increasing world population (Millennium Ecosystem Assessment 2005). Such worldwide losses of habitat are puffed up by increasingly large areas being cleared to meet the demand for biofuel, converting biodiverse lands into monocultures. Extensive tracts of tropical rainforest have been cleared to create oil-palm plantations for biodiesel in Indonesia and Malaysia (Dennis and Colfer 2006). Land to grow corn is increasing very fast to provide fodder for ethanol production in In the U.S. Midwest. More cerrado habitats are being replaced by soybean and sugar cane crops in Brazil (de Cerqueira Leite et al. 2009).

6.4.3 *Microalgae, a Possible Solution*

Microalgal can be ultimate most efficient source of biofuel production, in terms of land use and energy conversion (Chisti 2007; Lawton et al. 2016). Although the technical capacity to create large volumes of biofuel from microalgae have not yet been achieved (Ragauskas et al. 2006), we find this by far the most promising type of alternative, deserving of far greater attention and research. Among energy crops for which commercial-scale refining or demonstration projects are established, cellulosic ethanol and some biodiesels have shown strong energy returns, whereas much-less-developed alternative fuels derived from microalgae have astounding potential for high energy returns. Cellulosic ethanol is derived from grasses, crop and wood residues, and fastgrowing trees (such as poplar or willows) and typically yields >10 times as much energy as is needed to produce the fuel (Powlson et al. 2005).

Similarly, biodiesels have a high carbon content and return 2–6 times the energy used in production (Powlson et al. 2005; Hill et al. 2006).

With second-generation fuel-refining technology, cellulosic ethanol is expected to have much higher yields and consequently could have a much lower ecological footprint (Dien et al. 2003; Gray et al. 2006). The use of microalgae to produce biomass of high energy content has enormous potential for much higher energy yields and a much smaller ecological footprint (Sheehan et al. 1998; Kalscheuer et al. 2006; Chisti 2007). At present, microalgal biofuel are 4–10 times as expensive to produce as petroleum-derived fuels or other biodiesels (Chisti 2007). Nevertheless, only algal or microbial biofuel could be produced with a truly small ecological footprint because the space requirements for conventional crops or tree crops are 1–2 orders of magnitude greater. Even when grown in the least space-efficient manner (in large open ponds), only 200,000 ha would be needed to produce 1 quadrillion BTU from microalgae biodiesel (Sheehan et al. 1998), which is vastly less than the land area needed to produce a similar quantity of corn-derived ethanol (approximately 40 million ha) or soy biodiesel (approximately 20 million ha).

If microalgae were to reach its full potential, dedicating just 1.1% of U.S. cropland to microalgal production could replace half of the country's transportation fuel needs (Chisti 2007). Furthermore, many of the most promising species are diatoms and green algae that tolerate brackish or salt water and thus can be grown without use of increasingly scarce freshwater resources (Sheehan et al. 1998). Given the potential for much higher energy returns with microalgae, relative to other biofuel, this is an area that should be pursued actively.

In contrast to these higher potential yields, most estimates of energy returns from corn-derived ethanol show only a slight benefit, with a net energy balance of only 25%, or 1.25 times the energy needed to produce the fuel, because of the typically high inputs needed to grow the crop and the relatively low energy yield from this feedstock (Farrell et al. 2006; Hill et al. 2006). Thus, the current push to increase use of biofuel primarily through corn-based ethanol is clearly employing the least beneficial alternative fuel. Finally, biofuel may compete with arable land for growing food. In developing countries, this trade-off could result in social and economic problems. In the United States increased corn prices due to ethanol mandates have resulted in wide spread concern about impacts on livestock and other agricultural sectors, as well as on consumers.

6.5 Conclusion

Global water resources are more and more under pressure and the situation will be worse, as the demand for water accelerates due to expanding biofuel production. To acquire energy security and to meet sustainability in 2050. A sustainable biofuel production system must be built on the contemporary infrastructure and current technology, with the implementation of water saving innovative concepts. Obviously there is no single energy source that can sustainably fulfill future energy requirements entirely; however, biofuel are only choice that can meet the energy needs. Biofuel and bioproducts obtained from algal biomass will fulfill the future needs on industrial scale with the technologies for transforming biomass into biofuel that are economically feasible and environmentally friend. Mitigating the land use change impact, requires targeting biofuel production to degraded and abandoned cropland and rangeland; increasing crop yields, use of wastes, residues and compensatory offsite mitigation for residual direct and indirect impacts.

References

2030 Water Resources Group. 2009. Charting Our Water Future: Economic frameworks to inform decision-making http://www.2030waterresourcesgroup.com/water_full/Charting_Our_Water_Future_Final.pdf; Alexandratos N, Bruinsma J. World agriculture towards 2030/2050: the 2012 revision. ESA Working paper No. 12-03. Rome, FAO; 2012.

Adler PR, delGrosso SJ, Parton WJ. Life-cycle assessment of net greenhouse-gas flux for bioenergy cropping systems. Ecol Appl. 2007;17:675–91.

Alexandratos N, Bruinsma J. World agriculture towards 2030/2050: the 2012 revision (No. 12–03, p. 4). Rome, FAO: ESA Working paper; 2012.

Arshad M. Bioethanol: a sustainable and environment friendly solution for Pakistan. A Sci J COMSATS–Sci. Vision. 2010;16–7.

Arshad M, Ahmed S. Cogeneration through bagasse: a renewable strategy to meet the future energy needs. Renew Sust Energ Rev. 2016;54:732–7.

Arshad M, Khan ZM, Shah FA, Rajoka MI. Optimization of process variables for minimization of byproduct formation during fermentation of blackstrap molasses to ethanol at industrial scale. Lett Appl Microbiol. 2008;47:410–4.

Arshad M, Zia MA, Asghar M, Bhatti H. Improving bio-ethanol yield: using virginiamycin and sodium flouride at a Pakistani distillery. Afr J Biotechnol. 2011;10:11071.

Arshad M, Adil M, Sikandar A, Hussain T. Exploitation of meat industry by-products for biodiesel production: Pakistan's perspective. Pakistan J Life Soc Sci. 2014a;12:120–5.

Arshad M, Ahmed S, Zia MA, Rajoka MI. Kinetics and thermodynamics of ethanol production by *Saccharomyces cerevisiae* MLD10 using molasses. Appl Biochem Biotechnol. 2014b;172:2455–64.

Arshad M, Hussain T, Iqbal M, Abbas M. Enhanced ethanol production at commercial scale from molasses using high gravity technology by mutant *S. cerevisiae*. Brazilian J Microbiol. 2017. doi:10.1016/j.bjm.2017.02.003.

Bakker K, Morinville C. The governance dimensions of water security: a review. Phil Trans R Soc A. 2013;371:20130116.

Chartres C, Sood A. The water for food paradox. Aquatic Proc. 2013;1:3–19.

Chisti Y. Biodiesel from microalgae. Biotechnol Adv. 2007;25:294–306.

Conforti P. Looking ahead in world food and agriculture: perspectives to 2050. Food and Agriculture Organization of the United Nations (FAO). 2011.

Cosgrove WJ, Rijsberman FR. World water vision: making water everybody's business. Routledge; 2014 Mar 18.

Creutzig F, Goldschmidt JC, Lehmann P, Schmid E, von Blücher F, Breyer C, Fernandez B, Jakob M, Knopf B, Lohrey S, Susca T. Catching two European birds with one renewable stone: mitigating climate change and Eurozone crisis by an energy transition. Renew Sust Energ Rev. 2014;38:1015–28.

Damerau K, Patt AG, van Vliet OP. Water saving potentials and possible trade-offs for future food and energy supply. Glob Environ Change. 2016;39:15–25.

de Cerqueira Leite RC, Leal MR, Cortez LA, Griffin WM, Scandiffio MI. Can Brazil replace 5% of the 2025 gasoline world demand with ethanol? Energy. 2009;34:655–61.

Dennis RA, Colfer CP. Impacts of land use and fire on the loss and degradation of lowland forest in 1983–2000 in East Kutai District, East Kalimantan, Indonesia. Singapore J Trop Geogr. 2006;27:30–48.

Dien BS, Cotta MA, Jeffries TW. Bacteria engineered for fuel ethanol production: current status. Appl Microbiol Biotechnol. 2003;63:258–66.

Dominguez-Faus R, Powers SE, Burken JG, Alvarez PJ. The water footprint of biofuels: a drink or drive issue? Environ Sci Technol. 2009;43:3005–10.

EEA (European Environment Agency). How much bioenergy can Europe produce without harming the environment? Report 7/2006, ISSN 1725–9177. EEA, Copenhagen. 2006.

Eggert H, Greaker M. Promoting second generation biofuels: does the first generation pave the road? Energies. 2014;7:4430–45.

Fargione JE, Plevin RJ, Hill JD. The ecological impact of biofuels. Ann Rev Ecol Evol Syst. 2010;4:351–77.

Farrell AE, Plevin RJ, Turner BT, Jones AD, O'Hare M, Kammen DM. Ethanol can contribute to energy and environmental goals. Science. 2006;311:506–8.

Fingerman KR, Torn MS, O'Hare MH, Kammen DM. Accounting for the water impacts of ethanol production. Environ Res Lett. 2010;5:014020.

Gasparatos A, Stromberg P, Takeuchib K. Biofuels, ecosystem services and human wellbeing: putting biofuels in the ecosystem services narrative. Agric Ecosys Environ. 2011;142:111–28.

Gerbens-Leenes PW, Hoekstra AY, Meer TH. Water footprint of bio-energy and other primary energy carriers. 2008.

Graham RL, Liu W, English BC. The environmental benefits of cellulosic energy crops at a landscape scale. Environmental enhancement through agriculture: proceedings of a conference. Center for Agriculture, Food and Environment, Tufts University, Medford, Massachusetts. 1995.

Gray KA, Zhao L, Emptage M. Bioethanol. Curr Opin Chem Biol. 2006;10:141–6.

Griffing EM, Schauer RL, Rice CW. Life cycle assessment of fertilization of corn and corn–soybean rotations with swine manure and synthetic fertilizer in Iowa. J Environ Quality. 2014;43:709–22.

Groom MJ, Gray EM, Townsend PA. Biofuels and biodiversity: principles for creating better policies for biofuel production. Conserv Biol. 2008;22:602–9.

Hanjra MA, Qureshi ME. Global water crisis and future food security in an era of climate change. Food Policy. 2010;35:365–77.

Hardy L, Garrido A, Juana L. Evaluation of Spain's water-energy nexus. Int J Water Resour Dev. 2012;28:151–70.

Hertel TW, Golub A, Jones AD, O'Hare M, Plevin RJ, Kammen DM. Global land use and greenhouse gas emissions impacts of U.S. maize ethanol: estimating market-mediated responses. Bio Sci. 2010;60:223–31.

Hill J, Nelson E, Tilman D, Polasky S, Tiffany D. Environmental, economic, and energetic costs and benefits of biodiesel and ethanol biofuels. Proc Natl Acad Sci USA. 2006;103:11206–10.

Jiménez Cisneros BE, Oki T, Arnell NW, Benito G, Cogley JG, Doll P, Jiang T, Mwakalila SS. Fresh water resources. 2014:229–69.

Joly CA, Huntley BJ, Dale VH, Mace G, Muok B, Ravindranath NH. Biofuel impacts on biodiversity and ecosystem services. Scientific Committee on problems of the environment (SCOPE) rapid assessment process on bioenergy and sustainability. 2015;555–80.

Junginger M, Faaij A, Rosillo-Calle F, Wood J. The growing role of biofuels: opportunities, challenges, and pitfalls. Int Sugar J. 2006;108:615–29.

Kalscheuer R, St¨oveken T, Steinb¨uchel A. Engineered microorganisms for sustainable production of diesel fuel and other oleochemicals. Int Sugar J. 2006;109:1127.

Khan S, Khan MA, Hanjra MA, Mu J. Pathways to reduce the environmental footprints of water and energy inputs in food production. Food Policy. 2009;34:141–9.

Lal R. Soil and environmental implications of using crop residues as biofuel feedstock. Int Sugar J. 108:161–7.

Larson DF. Introducing water to an analysis of alternative food security policies in the Middle East and North Africa. Aquat Proc. 2013;1:30–43.

Lawton RJ, Cole AJ, Roberts DA, Paul NA, de Nys R. The industrial ecology of freshwater macroalgae for biomass applications. Algal Res. 2016.

Lele U, Klousia-Marquis M, Goswami S. Good governance for food, water and energy security. Aquat Procedia. 2013;1:44–63.

Macknick J, Newmark R, Heath G, Hallett KC. Operational water consumption and withdrawal factors for electricity generating technologies: a review of existing literature. Environ Res Lett. 2012;7:045802.

Mason L, Boyle T, Fyfe J, Smith T, Cordell D. National food waste data assessment: final report. Prepared for the Department of Sustainability, Environment, Water, Population and Communities. Sydney, Australia: Institute for Sustainable Futures, University of Technology. 2011.

Millennium Ecosystem Assessment. Ecosystems and human wellbeing: biodiversity synthesis. World Resources Institute, Washington, D.C. 2005.

Organisation for Economic Cooperation and Development (OECD). Environmental outlook to 2050: The consequences of inaction. Paris: OECD; 2012.

Parrish DJ, Fike JH. The biology and agronomy of switchgrass for biofuels. Criti Rev Plant Sci. 2005;24:423–59.

Perrone D, Murphy J, Hornberger GM. Gaining perspective on the water an energy nexus at the community scale. Environ Sci Technol. 2011;45:4228–34.

Powlson DS, Richie AB, Shield I. Biofuels and other approaches for decreasing fossil fuel emissions from agriculture. Ann Appl Biol. 2005;146:193–201.

Raboni M, Viotti P, Capodaglio AG. A comprehensive analysis of the current and future role of biofuels for transport in the European Union (EU). Revista Ambiente Agua. 2015;10:9–21.

Ragauskas AJ, Williams CK, Davison BH, Britovsek G, Cairney J, Eckert CA, Frederick WJ, Hallett JP, Leak DJ, Liotta CL, Mielenz JR. The path forward for biofuels and biomaterials. Science. 2006;311:484–9.

Rasul G. Food, water: and energy security in South Asia: a nexus perspective from the Hindu Kush Himalayan regions. Environ Sci Policy. 2014;39:35–48.

Rathmann R, Szklo A, Schaeffer R. Land use competition for production of food and liquid biofuels: an analysis of the arguments in the current debate. Renew Energ. 2010;35(1):14–22.

Ridoutt BG, Pfister S. A new water footprint calculation method integrating consumptive and degradative water use into a single stand-alone weighted indicator. Int J Life Cycle Ass. 2013;18:204–7.

Scott CA, Pierce SA, Pasqualetti MJ, Jones AL, Montz BE, Hoover JH. Policy and institutional dimensions of the water–energy nexus. Energ Policy. 2011;39:6622–30.

Searchinger T, Heimlich R, Houghton RA, Dong F, Elobeid A, Fabiosa J, Tokgoz S, Hayes D, Yu TH. Use of US croplands for biofuels increases greenhouse gases through emissions from land-use change. Science. 2008;319:1238–40.

Sheehan J, Dunahay T, Benemann J, Roessler P. A look back at the U.S. Department of Energy's aquatic species program-biodiesel from algae. Report to the Department of Energy. National Renewable Energy Laboratory, Golden, Colorado. 1998.

Sulser TB, Ringler C, Zhu T, Msangi M, Bryan E, Rosegrant MW. Green and blue water accounting in the Ganges and Nile basins: implications for food and agricultural policy. J Hydrol. 2010;384:276–91.

Tilman D, Hill J, Lehman C. Carbon negative biofuels from low-input, high-diversity grassland biomass. Science. 2006;314:1598–600.

Wang, M. Updated energy and greenhouse gas emissions results of fuel ethanol. In: The 15th international symposium on alcohol fuels. San Diego, California, USA, September, 2005.

Yang H, Zhou Y, Liu J. Land and water requirements of biofuel and implications for food supply and the environment in China. Energ Policy. 2009;37:1876–85.

MIX
Papier aus verantwortungsvollen Quellen
Paper from responsible sources
FSC® C105338

FSC
www.fsc.org

If you have any concerns about our products,
you can contact us on
ProductSafety@springernature.com

In case Publisher is established outside the EU,
the EU authorized representative is:
Springer Nature Customer Service Center GmbH
Europaplatz 3, 69115 Heidelberg, Germany

Printed by Libri Plureos GmbH
in Hamburg, Germany